Everyday Mathematics®

Student Math Journal 2

**The University of Chicago
School Mathematics Project**

McGraw Hill **Wright Group**

The McGraw·Hill Companies

UCSMP Elementary Materials Component

Max Bell, Director

Authors

Max Bell
John Bretzlauf
Amy Dillard
Robert Hartfield
Andy Isaacs

James McBride, Director
Kathleen Pitvorec
Peter Saecker
Robert Balfanz*
William Carroll*

Technical Art

Diana Barrie

*First Edition only

Photo Credits

Tony Freeman/Photo Edit p. 342, Cover: Bill Burlingham/Photography
Photo Collage: Herman Adler Design Group

Contributors

Tammy Belgrade, Diana Carry, Debra Dawson, Kevin Dorken, James Flanders, Laurel Hallman, Ann Hemwall, Elizabeth Homewood, Linda Klaric, Lee Kornhauser, Judy Korshak-Samuels, Deborah Arron Leslie, Joseph C. Liptak, Sharon McHugh, Janet M. Meyers, Susan Mieli, Donna Nowatzki, Mary O'Boyle, Julie Olson, William D. Pattison, Denise Porter, Loretta Rice, Diana Rivas, Michelle Schiminsky, Sheila Sconiers, Kevin J. Smith, Theresa Sparlin, Laura Sunseri, Kim Van Haitsma, John Wilson, Mary Wilson, Carl Zmola, Theresa Zmola

 This material is based upon work supported by the National Science Foundation under Grant No. ESI-9252984. Any opinions, findings, and conclusions or recommendations expressed in this material are those of the authors and do not necessarily reflect the views of the National Science Foundation.

Copyright © 2004 by Wright Group/McGraw-Hill.

Send all inquiries to:
Wright Group/McGraw-Hill
P.O. Box 812960
Chicago, IL 60681

Printed in the United States of America.

ISBN 0-07-600036-2

7 8 9 10 11 12 DBH 10 09 08 07 06 05

The McGraw-Hill Companies

Contents

Unit 7: Exponents and Negative Numbers

Unit 8: Fractions and Ratios

Unit 9: Coordinates, Area, Volume, and Capacity

Unit 10: Algebra Concepts and Skills

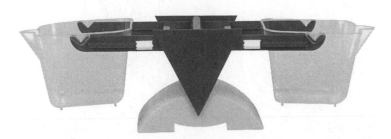

Unit 11: Volume

Unit 12: Probability, Ratios, and Rates

Exponents

Math Message

Which is correct, $4^3 = 12$ or $4^3 = 64$? Explain your answer. _____

_____

_____

Exponential Notation

In exponential notation, the **exponent** tells how many times the **base** is used as a factor. For example, $4^3 = 4 * 4 * 4 = 64$. The base is 4 and the exponent is 3.

1. Complete the table.

Exponential Notation	Base	Exponent	Repeated Factors	Product
5^4	5	4	5 * 5 * 5 * 5	625
	2	3		
			6 * 6 * 6 * 6	
			9 * 9	
			1 * 1 * 1 * 1 * 1 * 1 * 1	
	2			32

The Powers Key ⌃

2. Use your calculator to enter the keystrokes shown in the first column of the table.
   Record the calculator display in the second column.

   Study your results. What is the function of the ⌃ key?

   _____

   _____

   _____

   _____

Calculator Entry	Calculator Display
4 ⌃ 3 (Enter=)	
2 ⌃ 4 (Enter=)	
3 ⌃ 2 (Enter=)	
1 ⌃ 10 (Enter=)	
5 ⌃ 3 (Enter=)	

Use with Lesson 7.1.

Exponents (cont.)

Each problem below has a mistake. Find the mistake and tell what it is.
Then solve the problem.

3. $5^2 = 5 * 2 = 10$

Mistake: _____

_____

Correct solution: _____

4. $6^3 = 3 * 3 * 3 * 3 * 3 * 3 = 729$

Mistake: _____

_____

Correct solution: _____

5. $10^4 = 10 + 10 + 10 + 10 = 40$

Mistake: _____

_____

Correct solution: _____

Use your calculator to write the following numbers in standard notation.

6. $7 * 7 * 7 * 7 =$ _____

7. $15 * 15 * 15 * 15 =$ _____

8. $6^9 =$ _____

9. $5^8 =$ _____

10. $2^{12} =$ _____

11. 4 to the fifth power = _____

Write $<$, $>$, or $=$.

12. 10^2 _____ 2^{10}

13. 3^4 _____ 9^2

14. 1^2 _____ 1^5

15. 5^4 _____ 500

> **Reminder:**
> $>$ means *is greater than.*
> $<$ means *is less than.*

Use with Lesson 7.1.

Addition and Subtraction of Fractions

Find a common denominator. Then add or subtract.

1. $\dfrac{9}{10} - \dfrac{1}{5} =$ _____

2. $\dfrac{5}{9} - \dfrac{2}{5} =$ _____

3. $\dfrac{7}{12} + \dfrac{4}{5} =$ _____

4. $\dfrac{6}{15} - \dfrac{1}{10} =$ _____

5. $\begin{array}{r} \dfrac{1}{2} \\ -\ \dfrac{4}{9} \\ \hline \end{array}$

6. $\begin{array}{r} \dfrac{3}{7} \\ +\ \dfrac{5}{8} \\ \hline \end{array}$

Solve.

7. Regina is baking two different kinds of chocolate-chip cookies. One recipe calls for $\dfrac{1}{4}$ cup of chocolate chips. The other calls for $\dfrac{3}{4}$ cup of chocolate chips. How many cups of chocolate chips does she need in all?

   Write a number model: _____

   Solution: _____ cup

8. Roger found a long piece of heavy rope that was $24\dfrac{3}{4}$ feet long. It was a perfect rope for making jump ropes. If each jump rope is $8\dfrac{1}{4}$ feet long, how many can he make?

   _____ jump ropes

   Explain how you found your answer. _____

   _____

   _____

Math Boxes 7.1

1. Rewrite each fraction pair with common denominators.

a. $\frac{1}{3}$ and $\frac{1}{2}$ _____

b. $\frac{3}{4}$ and $\frac{2}{5}$ _____

c. $\frac{2}{8}$ and $\frac{9}{12}$ _____

SRB
65

2. Complete the table.

Fraction	Decimal	Percent
$\frac{7}{10}$		
$\frac{3}{8}$		
	$0.\overline{3}$	

SRB
83
89 90

3. Amanda collects dobsonflies (a pretty scary-looking creature, by all accounts). Below are the lengths, in millimeters, for the flies in her collection.

95, 107, 119, 103, 102, 91, 115, 120, 111, 114, 115, 107, 110, 107, 98, 112

a. Circle the stem-and-leaf plot below that represents this data.

Stems (100s and 10s)	Leaves (1s)
9	1 5 8
10	2 3 7 7 7
11	0 1 2 4 5 5 9
12	0

Stems (100s and 10s)	Leaves (1s)
9	1 5 8
10	2 3 7
11	0 1 2 4 5 9
12	0

Stems (100s and 10s)	Leaves (1s)
9	1 5 8 8 8
10	2 3 7 7 7
11	0 1 2 4 5 5 5
12	0

b. Find the following landmarks for the data.

Median: _____ Minimum: _____ Range: _____ Mode(s): _____

SRB
112 113

4. Divide. Show your work.

a. $843 \div 28 \rightarrow$ _____

b. $279 \div 17 \rightarrow$ _____

SRB
22–24

Use with Lesson 7.1.

Guides for Powers of 10

Study the place-value chart below.

	Millions			Thousands			Ones		
Billions	Hundred-millions	Ten-millions	Millions	Hundred-thousands	Ten-thousands	Thousands	Hundreds	Tens	Ones
10^9	10^8	10^7	10^6	10^5	10^4	10^3	10^2	10^1	10^0

In our place-value system, the powers of 10 are grouped into sets of three: ones, thousands, millions, billions, and so on. These groupings are helpful for working with large numbers. When we write large numbers, we separate these groups of three with commas.

We have prefixes for these groupings and for other important powers of 10. You know some of these prefixes from your work with the metric system. For example, the prefix *kilo-* in *kilometer* identifies a kilometer as 1,000 meters. It is helpful to memorize the exponential notation and the prefixes for one thousand, one million, one billion, and one trillion.

Use the place-value chart for large numbers and the prefixes chart to complete the following statements.

Prefixes	
tera-	trillion (10^{12})
giga-	billion (10^9)
mega-	million (10^6)
kilo-	thousand (10^3)
hecto-	hundred (10^2)
deca-	ten (10^1)
uni-	one (10^0)
deci-	tenth (10^{-1})
centi-	hundredth (10^{-2})
milli-	thousandth (10^{-3})
micro-	millionth (10^{-6})
nano-	billionth (10^{-9})

Example

1 kilogram equals $10^{\boxed{}}$ or one *thousand* grams.

1. The distance from Chicago to New Orleans is about 10^3 or one _____ miles.

2. A millionaire has at least $10^{\boxed{}}$ dollars.

3. A computer with 1 megabyte of RAM memory can hold approximately $10^{\boxed{}}$ or

   one _____ bytes of information.

4. A computer with a 1 gigabyte hard drive can store approximately $10^{\boxed{}}$ or

   one _____ bytes of information.

5. According to some scientists, the hearts of most mammals will beat about 10^9 or

   one _____ times in a lifetime.

Negative Powers of 10

Our base-ten place-value system works for decimals as well as for whole numbers.

Tens	Ones	.	Tenths	Hundredths	Thousandths
10s	1s	.	0.1s	0.01s	0.001s

Negative powers of 10 can be used to name decimal places.

Example $10^{-2} = \frac{1}{10^2} = \frac{1}{10 * 10} = \frac{1}{10} * \frac{1}{10} = 0.1 * 0.1 = 0.01$

Very small decimals can be hard to read in standard notation, so people often use number-and-word notation, exponential notation, or prefixes instead.

Guides for Small Numbers			
Number-and-Word Notation	**Exponential Notation**	**Standard Notation**	**Prefix**
1 tenth	$10^{-1} = \frac{1}{10}$	0.1	deci-
1 hundredth	$10^{-2} = \frac{1}{10 * 10}$	0.01	centi-
1 thousandth	$10^{-3} = \frac{1}{10 * 10 * 10}$	0.001	milli-
1 millionth	$10^{-6} = \frac{1}{10 * 10 * 10 * 10 * 10 * 10}$	0.000001	micro-
1 billionth	$10^{-9} = \frac{1}{10 * 10 * 10 * 10 * 10 * 10 * 10 * 10 * 10}$	0.000000001	nano-
1 trillionth	$10^{-12} = \frac{1}{10 * 10 * 10 * 10 * 10 * 10 * 10 * 10 * 10 * 10 * 10 * 10}$	0.000000000001	pico-

Use the table above to complete the following statements.

1. A fly can beat its wings once every 10^{-3} seconds, or once every one thousandth

of a second. This is a _____ second.

2. Earth travels around the sun at a speed of about 1 inch per microsecond.

This is $10^{\boxed{}}$ second, or a _____ of a second.

3. Electricity can travel one foot in a nanosecond, or one _____ of a

second. This is $10^{\boxed{}}$ second.

4. In $10^{\boxed{}}$ second, or one picosecond, an air molecule can spin once.

This is a _____ of a second.

Math Boxes 7.2

1. Multiply. Use the partial-products algorithm.

   a. 87
        * 65

   b. 39
        * 24

   c. 99
        * 26

SRB 19

2. Tell whether each number is prime or composite.

   a. Number of hours in $\frac{1}{3}$ of a day   _____

   b. Number of minutes in $\frac{1}{12}$ of an hour   _____

   c. Number of weeks in $\frac{1}{4}$ of a year   _____

   d. Number of months in $\frac{2}{3}$ of a year   _____

   e. Number of days in $\frac{3}{7}$ of a week   _____

SRB 12 74

3. What is the measure of angle T?

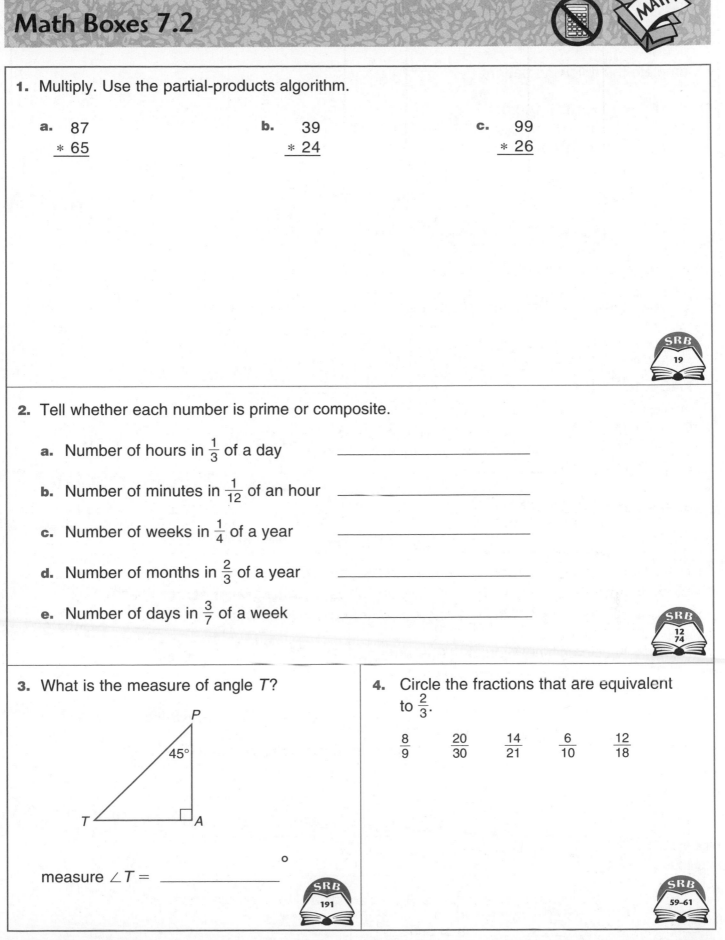

   measure $\angle T =$ _____ °

SRB 191

4. Circle the fractions that are equivalent to $\frac{2}{3}$.

   $\frac{8}{9}$ $\frac{20}{30}$ $\frac{14}{21}$ $\frac{6}{10}$ $\frac{12}{18}$

SRB 59–61

Scientific Notation

Complete the following pattern.

1. $10^2 = 10 * 10 = 100$

2. $10^3 = 10 * 10 * 10 =$ _____

3. $10^4 =$ _____ $=$ _____

4. $10^5 =$ _____ $=$ _____

5. $10^6 =$ _____ $=$ _____

Use your answers in Problems 1–5 to help you complete the following.

6. $2 * 10^2 = 2 * 100 = 200$

7. $3 * 10^3 = 3 *$ _____ $=$ _____

8. $4 * 10^4 =$ _____ $*$ _____ $=$ _____

9. $6 * 10^5 =$ _____ $*$ _____ $=$ _____

10. $8 * 10^6 =$ _____ $*$ _____ $=$ _____

___

Numbers written as the product of a number and a power of 10 are said to be in
scientific notation. Scientific notation is a useful way of writing large or small numbers.
Many calculators display numbers one billion or larger with scientific notation.

Example In scientific notation, 4,000 is written as $4 * 10^3$.
It is read as "four times ten to the third power."

Write each of the following in standard notation and number-and-word notation.

	Standard Notation	**Number-and-Word Notation**
11. $5 * 10^3 =$	_____	_____
12. $7 * 10^2 =$	_____	_____
13. $2 * 10^4 =$	_____	_____
14. $5 * 10^6 =$	_____	_____

Use with Lesson 7.3.

Math Boxes 7.3

1. **a.** What is the perimeter of the rectangle?

_____

12 units

8 units

b. What is the area?

_____

SRB
170

2. **a.** Find an object in the room that is about 15 centimeters long.

_____

b. Find an object in the room that is about 3 inches long.

_____

SRB
169

3. **a.** Draw a circle that has a diameter of 4 centimeters.

b. The radius of the circle is

_____ cm.

SRB
143 154

4. Use a calculator to rename each of the following in standard notation.

a. 3^{10} = _____

b. 8^4 = _____

c. 4^8 = _____

d. 5^7 = _____

e. 9^8 = _____

SRB
6 244

5. Solve. Do not use a calculator.

a. 287 + 395 = _____

b. 712 + 504 = _____

c. 776 + _____ = 1,943

d. 2,080 = 948 + _____

e. _____ + 286 = 345

SRB
13–17

History of the Earth

Geologists, anthropologists, paleontologists, and other scholars often estimate when important events occurred in the history of the Earth. For example, when did dinosaurs become extinct? When did the Rocky Mountains develop? The estimates are very broad, partly because events like these lasted for many years, and partly because dating methods cannot precisely pinpoint exact times so long ago.

Scientists base their estimates on the geological record—rocks, fossils, and other clues—and on the bones and tools left by humans long ago. Below is a list of events prepared by one group of scientists. All the data are approximations, and different estimates are given by other scientists.

Use the place-value chart on the next page to help you write, in standard notation, how long ago the events below took place.

Example Earth was formed about $5 * 10^9$ years ago. Find 10^9 on the place-value chart and write 5 beneath it, followed by zeros in the cells to the right. Then use the chart to help you read the number: $5 * 10^9 = 5$ billion.

What happened:	**Some scientists say it happened about this many years ago:**
1. Earth was formed.	$5 * 10^9$ years
2. The first signs of life (bacteria cells) appeared.	$4 * 10^9$
3. Fish appeared.	$4 * 10^8$
4. Forests, swamps, insects, and reptiles appeared.	$3 * 10^8$
5. Tyrannosaurus Rex lived; modern trees appeared.	$1 * 10^8$
6. The first known human-like primates appeared.	$6 * 10^6$
7. Woolly mammoths and other large ice-age mammals appeared.	$8 * 10^5$
8. Humans first moved from Asia to North America.	$2 * 10^4$

Challenge

9. The first dinosaurs appeared; the Appalachian Mountains formed.	$2.5 * 10^8$
10. Dinosaurs became extinct.	$6.5 * 10^7$

Source: The Handy Science Answer Book

History of the Earth (cont.)

	Billion	100 M	10 M	Million	100 Th	10 Th	Thousand	100	10	One
	10^9	10^8	10^7	10^6	10^5	10^4	10^3	10^2	10^1	10^0
1.										
2.										
3.										
4.										
5.										
6.										
7.										
8.										
9.										
10.										

Reminder: Powers are calculated before other factors are multiplied. $5.5 * 10^4 = 5.5 * 10{,}000 = 55{,}000$.

That's a Yotta Numbers

Prefixes for very large and very small numbers, such as *tera-* (10^{12}) and *pico-* (10^{-12}), were adopted by an international scientific and mathematics committee in 1960. Since then, scientists and mathematicians have routinely worked with still larger and smaller numbers and have updated the list of prefixes.

The most recent adoption was yotta- (10^{24}) in 1991. *Yotta-* is based on the Latin word for *eight,* because 1 septillion is equal to 1,000 to the eighth power ($1{,}000^8$). Can you think of a prefix for 1 octillion and 1 nonillion?

Number	Exponential Notation	Prefix
1 quadrillion	10^{15}	peta-
1 quintillion	10^{18}	exa-
1 sextillion	10^{21}	zetta-
1 septillion	10^{24}	yotta-
1 octillion	10^{27}	
1 nonillion	10^{30}	

History of the Earth (cont.)

Work with a partner to answer the following questions.

11. According to the estimates of scientists, about how many years passed from the formation of Earth until the first signs of life?

_____

12. About how many years passed between the appearance of the first fish and the appearance of forests and swamps?

_____

13. Make up and answer one or two questions of your own about data in the table of

Earth's history. _____

_____

_____

_____

_____

_____

_____

_____

_____

Challenge

14. According to the geological record, about how long did dinosaurs roam on Earth?

_____

Math Boxes 7.4

1. Rewrite each fraction pair with common denominators.

   a. $\frac{2}{3}$ and $\frac{3}{5}$ _____

   b. $\frac{3}{7}$ and $\frac{9}{10}$ _____

   c. $\frac{3}{8}$ and $\frac{18}{24}$ _____

2. Complete the table.

Fraction	Decimal	Percent
$\frac{2}{3}$		
	0.95	
		43%
$\frac{3}{5}$		
	0.8	

3. a. Make a stem-and-leaf plot of the hand-span measures in Ms. Grip's fifth grade class.

   163, 179, 170, 165, 182, 157, 154, 165, 170, 175, 162, 185, 158, 158, 170, 165, 162, 154

   b. Find the following landmarks for the data.

   Median: _____ Minimum: _____

   Range: _____ Mode(s): _____

4. Divide. Show your work.

   a. $21\overline{)493}$ _____

   b. $35\overline{)623}$ _____

Parentheses and Number Stories

Math Message

1. Make a true sentence by filling in the missing number.

    a. $7.3 - (2.2 + 1.1) =$ _____

    b. $(7.3 - 2.2) + 1.1 =$ _____

    c. $2.0 * (7.5 + 1.5) =$ _____

    d. $(2.0 * 7.5) + 1.5 =$ _____

2. Solve the following problem to get as many different answers as possible.
    Write a number sentence for each way.

    $$6 * 4 - 2 / 2 = ?$$

    _____

    _____

    _____

    _____

    _____

Match each number story with the expression that fits that story.

3. **Story 1**

    Tom had 4 cans of soda.
    He went shopping and bought
    3 six-packs of soda cans.

    Story 2

    Tom had 4 six-packs of soda cans.
    He went shopping and bought 3 more
    six-packs of soda cans.

    Tom's total number of soda cans:

    $(4 + 3) * 6$

    $4 + (3 * 6)$

Use with Lesson 7.4.

Parentheses and Number Stories (cont.)

4. Story 1

Alice ate 3 cookies before going to a party.
At the party, Alice and 4 friends ate equal
shares of 45 cookies.

Number of cookies Alice ate:

3 + (45 / 5)

Story 2

There was a full bag with 45 cookies,
and an opened bag with 3 cookies.
Alice and 4 friends ate equal shares
of all these cookies.

(3 + 45) / 5

5. Story 1

Mr. Chung baked 5 batches of cookies.
Each of the first 4 batches contained
15 cookies. The final batch contained
only 5 cookies.

Number of cookies baked:

15 * (4 + 5)

Story 2

In the morning, Mr. Chung baked
4 batches of 15 cookies each. In the
afternoon, he baked 5 more batches
of 15 cookies each.

(15 * 4) + 5

6. A grocery store received a shipment of 120 cases of apple juice. Each
case contained 4 six-packs of cans. After inspection, the store found that
9 cans were damaged.

Write an expression that represents the number of undamaged cans.

_____

Order of Operations

Use the rules of order of operations to complete these number sentences.

1. $100 + 500 / 2 =$ _____

2. $24 / 6 + 3 * 2 =$ _____

3. $2 * 4^2 =$ _____

4. $25 - 10 + 5 * 2 + 100 / 20 =$ _____

5. $24 / 6 / 2 + 12 - 3 * 2 =$ _____

Insert parentheses in each problem below to get as many different answers as you can.
The first one is done as an example.

6. $5 + 4 * 9 =$ $(5 + 4) * 9 = 81$ $5 + (4 * 9) = 41$

7. $4 * 3 + 10 =$ _____

8. $6 * 4 / 2 =$ _____

9. $10 - 6 - 4 =$ _____

First, solve these problems by hand. Then solve them with your calculator.

	By Hand	Calculator
10.	$5 + 3 * 6 =$ _____	$5 + 3 * 6 =$ _____
11.	$3 * 6 + 5 =$ _____	$3 * 6 + 5 =$ _____
12.	$36 - 18 / 6 =$ _____	$36 - 18 / 6 =$ _____
13.	$44 - 6 * 5 =$ _____	$44 - 6 * 5 =$ _____

14. a. Does your calculator obey the correct order of operations? _____

b. If your calculator obeys the correct order of operations, how do you know?

_____

_____

c. If your calculator doesn't obey the correct order of operations, then what order

does it use? _____

_____

Use with Lesson 7.5.

American Tour: Inequalities

Use the American Tour section of your *Student Reference Book* to make
comparisons of population, geographic area, or other data. Use > or < to write an
inequality for each comparison.

Symbol	Meaning
>	*is greater than*
<	*is less than*

	Comparison	Inequality
1.		
2.		
3.		
4.		
5.		
6.		
7.		

Use with Lesson 7.5.

Math Boxes 7.5

1. Multiply. Use the partial-products algorithm.

   a.   43
            * 78

   b.   19
            * 86

   c.   79
            * 42

2. Tell whether each number is prime or composite.

   a. Number of millimeters in 2.9 centimeters _____

   b. Number of inches in $1\frac{1}{2}$ yards   _____

   c. Number of centimeters in 0.35 meter   _____

   d. Number of inches in $\frac{5}{6}$ foot   _____

   e. Number of feet in $4\frac{1}{3}$ yards   _____

3. What is the measure of angle *R*?

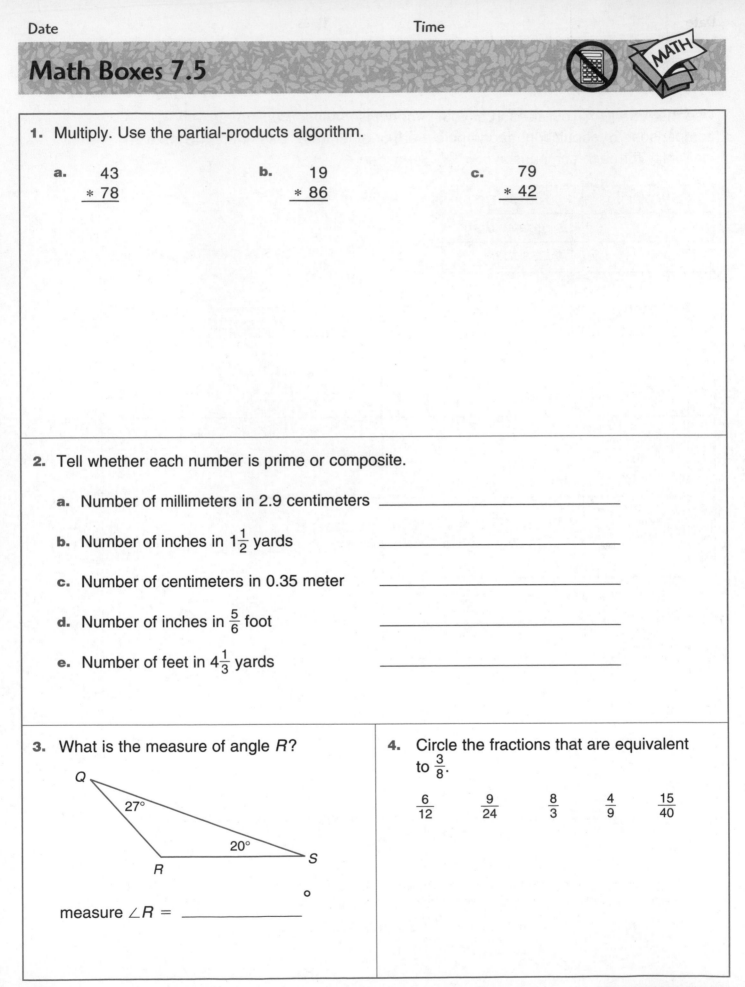

   measure ∠*R* = _____

4. Circle the fractions that are equivalent to $\frac{3}{8}$.

   $\frac{6}{12}$ $\frac{9}{24}$ $\frac{8}{3}$ $\frac{4}{9}$ $\frac{15}{40}$

Positive and Negative Numbers on a Number Line

1. Graph each of the following bicycle race events on the number line below. Label each event with its letter. (*Hint:* Zero on the number line stands for the starting time of the race.)

   A Check in 5 minutes before the race starts.

   B Change gears 30 seconds after the race starts.

   C Get on the bicycle 30 seconds before the race starts.

   D The winner finishes at 6 minutes, 45 seconds.

   E Complete the first lap 3 minutes, 15 seconds after the race starts.

   F Check the tires 2 minutes before the race starts.

Minutes

2. Mr. Pima's class planned a raffle. Five students were asked to sell raffle tickets. The goal for each student was $50 in ticket sales. The table below shows how well each of the five students did. Complete the table. Then graph the amounts from the last column on the number line below the table. Label each amount with that student's letter.

Student	Ticket Sales	Amount That Ticket Sales Were Above or Below Goal
A	$5.50 short of goal	–$5.50
B	Met goal exactly	
C	Exceeded goal by $1.75	
D	Sold $41.75	
E	Sold $53.25	

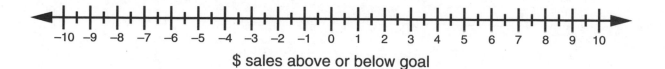

$ sales above or below goal

Comparing and Ordering Numbers

For any pair of numbers on the number line, the number to the left is less than the number to the right.

$$\xleftarrow{\quad} \; -10 \; -9 \; -8 \; -7 \; -6 \; -5 \; -4 \; -3 \; -2 \; -1 \; 0 \; 1 \; 2 \; 3 \; 4 \; 5 \; 6 \; 7 \; 8 \; 9 \; 10 \; \xrightarrow{\quad}$$

−10 is less than −5, because −10 is to the left of −5.
We use the < (less than) symbol to write −10 < −5.

+10 is greater than +5, because +10 is to the right of +5.
We use the > (greater than) symbol to write +10 > +5.

> *Reminder:*
> When writing the > or < symbol, be sure the arrow tip points to the smaller number.

Write > or <.

1. −5 _____ 5

2. 10 _____ −10

3. −10 _____ 0

4. 14 _____ 7

5. −14 _____ −7

6. 0 _____ $-6\frac{1}{2}$

Answer the following.

7. What is the value of π to two decimal places? _____

8. $-\pi =$ _____

List the numbers in order from least to greatest.

9. −10, 14, −100, $\frac{8}{2}$, −17, 0 _____

10. −0.5, 0, −4, −π, −4.5 _____

Answer the following.

11. Name four positive numbers less than π. _____

12. Name four negative numbers greater than −π. _____

        Use with Lesson 7.6.

Review and Practice with Parentheses

Solve.

1. $(4 + 5) / 3 =$ _____

2. $(3 + 2) * (4 - 2) =$ _____

3. $((3 + 2) * (4 - 2)) / 2 =$ _____

4. $5 * ((5 + 5) * (5 + 5)) =$ _____

5. _____ $= 32 / (16 / 2)$

6. _____ $= (32 / 16) / 2$

7. $(6.5 + 8.3) / (3 - 1) =$ _____

Make each sentence true by inserting parentheses.

8. $18 \ - \ 11 \ + \ 3 \ = \ 10$

9. $18 \ - \ 11 \ + \ 3 \ = \ 4$

10. $14 \ - \ 7 \ + \ 5 \ + \ 1 \ = \ 13$

11. $14 \ - \ 7 \ + \ 5 \ + \ 1 \ = \ 1$

12. $14 \ - \ 7 \ + \ 5 \ + \ 1 \ = \ 3$

13. $100 \ = \ 15 \ + \ 10 \ * \ 4$

14. $4 \ = \ 24 \ / \ 4 \ + \ 2$

15. $8 \ = \ 24 \ / \ 4 \ + \ 2$

16. $10 \ - \ 4 \ / \ 2 \ * \ 3 \ = \ 24$

17. $10 \ - \ 4 \ / \ 2 \ * \ 3 \ = \ 1$

Use with Lesson 7.6.

Math Boxes 7.6

---

1. a. A rectangle has a perimeter of 12 cm. One side is 4 cm long. Draw the rectangle.

b. What is the area of the rectangle you drew? _____

2. a. Find an object in the room that is about 10 inches long.

_____

b. Find an object in the room that is about 10 centimeters long.

_____

---

3. Find the radius and diameter of the circle.

Radius = _____
                    (unit)

Diameter = _____
                      (unit)

4. Use a calculator to rename each of the following in standard notation.

a. $7^3 =$ _____

b. $9^5 =$ _____

c. $4^5 =$ _____

d. $6^8 =$ _____

e. $3^7 =$ _____

---

5. Solve. Do not use a calculator.

a.	b.	c.	d.	e.
243	385	1,006	6,463	513
+ 477	+ 948	− 597	+ 2,099	− 475

Use with Lesson 7.6.

500

Math Message

The game *500* is a bat-and-ball game for two or more players. One player hits balls to the other players. The other players score points by catching the hit balls. Scoring is shown at the right.

Catch	Points
fly	100
one bounce	75
two bounces	50
grounder	25

If a player drops a ball, then the points are subtracted. For example, if a player tries to catch a fly ball and drops it, then 100 points are subtracted from the player's score.

The first player to reach 500 points becomes the next batter and the game starts over.

Sometimes, players have to go "in the hole." This happens when they miss a catch worth more points than they have. For example, if the first hit of the game is a fly and a player misses it, that player is 100 points "in the hole."

1. Complete the following table for a game of *500*.

Action	Points Scored	Total Score
caught grounder	+25	25
missed fly	−100	75 in the hole
caught two-bouncer	+50	
caught fly		
missed fly		
missed one-bouncer		
missed fly		
caught fly		
caught fly		
caught fly		
missed one-bouncer		
caught fly		

2. Evan was 125 in the hole. How might he have gotten that score? _____

_____

_____

Using Counters to Show an Account Balance

Use *Math Masters,* page 96. Shade the ⊞ squares with a regular pencil and the ⊟ squares with a red pencil or crayon. Then cut out the squares.

- Each ⊞ counter represents $1 of cash on hand.

- Each ⊟ counter represents a $1 debt, or $1 that is owed.

Your **account balance** is the amount of money that you have or that you owe.
If you have money in your account, your balance is **"in the black."**
If you owe money, your account is **"in the red."**

1. Suppose you have this set of counters.   ⊞ ⊞ ⊞ ⊞ ⊞ ⊟ ⊟ ⊟

   a. What is your account balance? _____

   b. Are you "in the red" or "in the black"? _____

2. Use ⊞ and ⊟ counters to show an account with a balance of +$5. Draw a picture of the counters below.

3. Use ⊞ and ⊟ counters to show an account with a balance of −$8. Draw a picture of the counters below.

4. Use ⊞ and ⊟ counters to show an account with a balance of $0. Draw a picture of the counters below.

Addition of Positive and Negative Numbers

Use your counters to help you solve these problems. Draw $\boxed{+}$ and $\boxed{-}$ counters to show how you solved each problem.

1. $+8 + (-2) =$ _____

2. $-4 + (-5) =$ _____

3. $-3 + (+7) =$ _____

Solve these addition problems.

4. $50 + (-30) =$ _____   5. _____ $= -50 + 30$

6. $-16 + 10 =$ _____   7. _____ $= 16 + (-10)$

8. $-9 + (-20) =$ _____   9. _____ $= -15 + 15$

10. $27 + (-18) =$ _____   11. _____ $= -43 + (-62)$

12. $-17 + (-17) =$ _____   13. _____ $= -55 + 32$

Challenge

14. The temperature at sunset was 13°C. During the night, the temperature dropped 22°C. Write a number model and figure out the temperature at sunrise the next morning.

    Number model: _____

    Answer: _____

Math Boxes 7.7

1. Write each fraction as a mixed number or a whole number.

a. _____ $= \dfrac{38}{3}$

b. _____ $= \dfrac{83}{7}$

c. _____ $= \dfrac{42}{6}$

d. _____ $= \dfrac{28}{11}$

e. _____ $= \dfrac{47}{12}$

SRB 62 63

2. Subtract. (*Hint:* Use a number line to help you.)

a. $25 - 25 =$ _____

b. $25 - 27 =$ _____

c. $15 - 18 =$ _____

d. $46 - 50 =$ _____

e. $38 - 82 =$ _____

SRB 92

3. Write each numeral in number-and-word notation.

a. 43,000,000 _____

b. 607,000 _____

c. 3,000,000,000 _____

d. 72,000 _____

SRB 4

4. Round each number to the nearest thousand.

a. 7,091 _____

b. 35,658 _____

c. 829,543 _____

d. 105,799 _____

e. 372,372 _____

SRB 227

5. Complete the "What's My Rule?" table and state the rule.

Rule

◯	▢
100	
9	0.9
50	5
	1.5
	0.5

SRB 215 216

6. A person breathes an average of 12 to 15 times per minute. At this rate, about how many breaths might a person take in a day? _____

Explain how you got your answer.

_____

_____

_____

SRB 221–223

Use with Lesson 7.7.

Finding Balances

In the following problems, use your $1 cash cards as ⊞ counters and your $1 debt cards as ⊟ counters. The balance is the total value of the combined ⊞ and ⊟ counters.

Draw a picture of the ⊞ and ⊟ counters to show how you found each balance.

1. You have 3 ⊟ counters. Add 6 ⊞ counters.

   Balance = _____ counters

2. You have 5 ⊞ counters. Add 7 ⊟ counters.

   Balance = _____ counters

3. You have 5 ⊞ counters. Add 5 ⊟ counters.

   Balance = _____ counters

4. Show a balance of –7 using 15 of your ⊞ and ⊟ counters.

5. You have 7 ⊟ counters. Take away 4 ⊟ counters.

   Balance = _____ counters

6. You have 7 ⊞ counters. Take away 4 ⊟ counters.

   Balance = _____ counters

7. You have 7 ⊟ counters. Take away 4 ⊞ counters.

   Balance = _____ counters

Adding and Subtracting Numbers

You and your partner combine your ⊞ and ⊟ counters. Use the counters to help you solve the problems.

1.

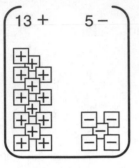

Balance = _____
If 4 ⊟ counters are subtracted from the container, what is the new balance?

New balance = _____

Number model: _____

2.

13 +   5 −

Balance = _____
If 4 ⊞ counters are added to the container, what is the new balance?

New balance = _____

Number model: _____

3.

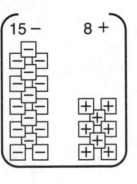

Balance = _____
If 3 ⊞ counters are subtracted from the container, what is the new balance?

New balance = _____

Number model: _____

4.

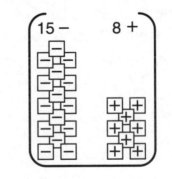

Balance = _____
If 3 ⊟ counters are added to the container, what is the new balance?

New balance = _____

Number model: _____

Use with Lesson 7.8.

Adding and Subtracting Numbers (cont.)

5.

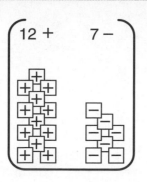

Balance = _____

If 6 ⊟ counters are subtracted
from the container, what is the
new balance?

New balance = _____

Number model: _____

6.

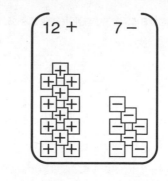

Balance = _____

If 6 ⊞ counters are added to
the container, what is the
new balance?

New balance = _____

Number model: _____

7.

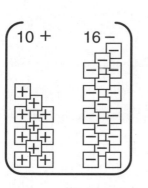

Balance = _____

If 2 ⊟ counters are subtracted
from the container, what is the
new balance?

New balance = _____

Number model: _____

8.

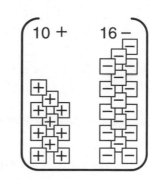

Balance = _____

If 2 ⊞ counters are added to
the container, what is the
new balance?

New balance = _____

Number model: _____

9. Write a rule for subtracting positive and negative numbers.

_____

_____

_____

Subtraction Problems

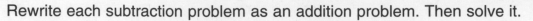

Rewrite each subtraction problem as an addition problem. Then solve it.

1. $100 - 45 =$ $\underline{\quad 100 + (-45) \quad}$ = _____

2. $-100 - 45 =$ _____ = _____

3. $160 - (-80) =$ _____ = _____

4. $9 - (-2) =$ _____ = _____

5. $-4 - (-2) =$ _____ = _____

6. $-15 - (-30) =$ _____ = _____

7. $8 - 10 =$ _____ = _____

8. $-20 - (-7) =$ _____ = _____

9. $\pi - (-\pi) =$ _____ = _____

10. $0 - (-6.1) =$ _____ = _____

Challenge

11. The Healthy Delights Candy Company specializes in candy that is wholesome and good for you. Unfortunately, they have been losing money for several years. During the year 2000, they lost $12 million, ending the year with a total debt of $23 million.

    a. What was Healthy Delights' total debt at the beginning of 2000? _____

    b. Write a number model that fits this problem. _____

12. In 2001, Healthy Delights is expecting to lose $8 million.

    a. What will Healthy Delights' total debt be at the end of 2001? _____

    b. Write a number model that fits this problem. _____

Use with Lesson 7.8.

Math Boxes 7.8

1. Find the whole set.

   a. 4 is $\frac{1}{8}$ of the set. _____

   b. 4 is $\frac{2}{5}$ of the set. _____

   c. 9 is $\frac{3}{7}$ of the set. _____

   d. 5 is $\frac{1}{3}$ of the set. _____

   e. 12 is $\frac{3}{8}$ of the set. _____

SRB 75

2. Make true sentences by inserting parentheses.

   a. $5 * 4 - 2 = 10$

   b. $25 + 8 * 7 = 81$

   c. $36 / 6 - 5 = 36$

   d. $45 / 9 + 6 = 11$

   e. $45 / 9 + 6 = 3$

SRB 203 204

3. Make a circle graph of the survey results.

Favorite After-School Activity	
Activity	**Students**
Eat Snack	18%
Visit Friends	35%
Watch TV	22%
Read	10%
Play Outside	15%

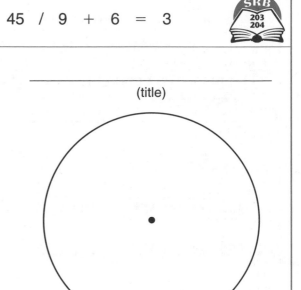

_____
(title)

SRB 120 121

4. Add. Use fraction sticks to help you.

   a. $\frac{1}{4} + \frac{2}{4} =$ _____

   b. $\frac{3}{8} + \frac{1}{4} =$ _____

   c. $\frac{1}{2} + \frac{1}{8} =$ _____

   d. $\frac{2}{3} + \frac{1}{6} =$ _____

   e. $\frac{2}{6} + \frac{2}{6} =$ _____

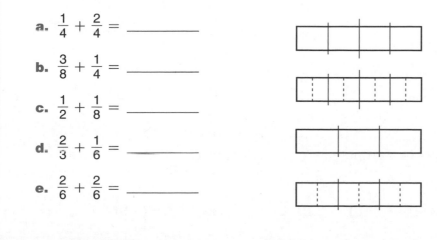

Date _____   Time _____

Addition and Subtraction on a Slide Rule

Math Message Find each sum or difference.

1. $13 - (+10) =$ _____   **2.** $13 - 10 =$ _____   **3.** $13 + (-10) =$ _____

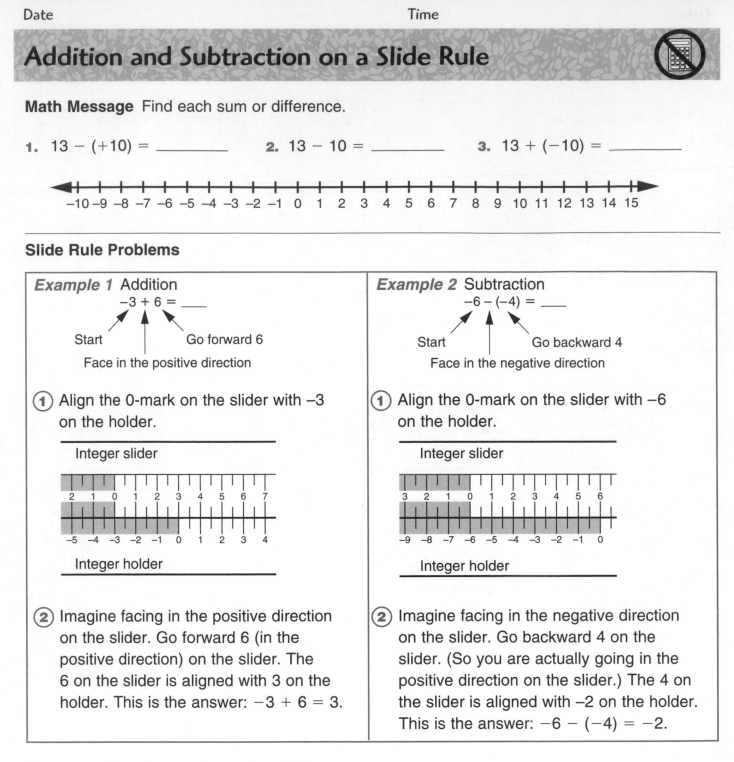

Slide Rule Problems

Example 1 Addition
$-3 + 6 =$ ___

Start — Go forward 6
Face in the positive direction

① Align the 0-mark on the slider with –3 on the holder.

Integer slider

Integer holder

② Imagine facing in the positive direction on the slider. Go forward 6 (in the positive direction) on the slider. The 6 on the slider is aligned with 3 on the holder. This is the answer: $-3 + 6 = 3$.

Example 2 Subtraction
$-6 - (-4) =$ ___

Start — Go backward 4
Face in the negative direction

① Align the 0-mark on the slider with –6 on the holder.

Integer slider

Integer holder

② Imagine facing in the negative direction on the slider. Go backward 4 on the slider. (So you are actually going in the positive direction on the slider.) The 4 on the slider is aligned with –2 on the holder. This is the answer: $-6 - (-4) = -2$.

Use your slide rule to solve each problem.

4. $12 - 17 =$ _____   **5.** $12 + (-17) =$ _____   **6.** $10 - (-4) =$ _____

7. $10 + 4 =$ _____   **8.** $-10 - (-5) =$ _____   **9.** $6 - 13 =$ _____

10. $-2 + (-13) =$ _____   **11.** $-5 - 10 =$ _____   **12.** $-8 + 8 =$ _____

13. $-8 - 8 =$ _____   **14.** $-8 + (-8) =$ _____   **15.** $-8 - (-8) =$ _____

Use with Lesson 7.9.

Using a Ruler

1. Mark each of these lengths on the ruler shown below. Write the letter above your mark. Point *A* has been done for you.

    A: $2\frac{1}{16}$ in.   *B:* $4\frac{3}{8}$ in.   *C:* $3\frac{3}{4}$ in.   *D:* $1\frac{7}{16}$ in.   *E:* $2\frac{4}{8}$ in.

A

0   1   2   3   4   5   6
INCHES

2. Measure the following line segments to the nearest $\frac{1}{16}$ of an inch.

    a. _____

         _____ in.

    b. _____

         _____ in.

    c. _____

         _____ in.

    d. _____

         _____ in.

3. Draw a line segment that is $4\frac{3}{10}$ inches long.

4. Draw a line segment that is $3\frac{1}{2}$ inches long.

5. Complete these ruler puzzles.

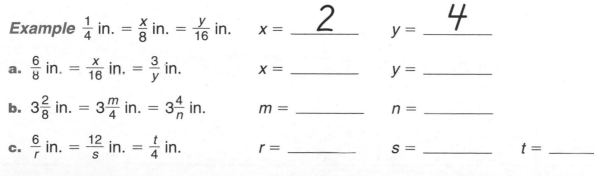

    Example $\frac{1}{4}$ in. $= \frac{x}{8}$ in. $= \frac{y}{16}$ in. $x =$ ___2___ $y =$ ___4___

    a. $\frac{6}{8}$ in. $= \frac{x}{16}$ in. $= \frac{3}{y}$ in. $x =$ _____ $y =$ _____

    b. $3\frac{2}{8}$ in. $= 3\frac{m}{4}$ in. $= 3\frac{4}{n}$ in. $m =$ _____ $n =$ _____

    c. $\frac{6}{r}$ in. $= \frac{12}{s}$ in. $= \frac{t}{4}$ in. $r =$ _____ $s =$ _____ $t =$ _____

Math Boxes 7.9

1. Write each mixed number as a fraction.

a. _____ $= 3\frac{4}{7}$

b. _____ $= 5\frac{2}{3}$

c. _____ $= 6\frac{8}{9}$

d. _____ $= 4\frac{12}{9}$

e. _____ $= 8\frac{6}{4}$

2. Subtract. (*Hint:* Use a number line to help you).

a. $32 - 38 =$ _____

b. $14 - 21 =$ _____

c. $84 - 85 =$ _____

d. $36 - 52 =$ _____

e. $40 - 73 =$ _____

3. Write each numeral in number-and-word notation.

a. 56,000,000 _____

b. 423,000 _____

c. 18,000,000,000 _____

d. 9,500,000 _____

4. Round each number to the nearest tenth.

a. 45.06 _____

b. 29.95 _____

c. 1.005 _____

d. 7.98 _____

e. 5.76 _____

5. Complete the "What's My Rule?" table and state the rule.

Rule		◯	☐
		28	7
		16	
		1	
			5
		0	

6. Marcus had $5.00 to spend on lunch. He bought a hot dog for $1.75 and some french fries for $0.69. How much money did he have left for dessert?

_____

Use with Lesson 7.9.

Entering Negative Numbers on a Calculator

Math Message

1. Press the ⊖ key, then 3, and then (Enter) . What number is shown in the display?

    _____

2. Repeat the steps in Problem 1 with other numbers.

3. **a.** What is the opposite of 5? _____
    Enter the opposite of 5 in the calculator.

    b. What is the opposite of the opposite of 5? _____
    Enter this number in the calculator, using the ⊖ key.

4. What does the ⊖ key do? _____

    _____

    _____

    _____

Addition and Subtraction Using a Calculator

Use your calculator to solve each problem. Record how you did it.

Example   **Calculator Entry**

$12 + (-17) =$ -5 12 ⊞ ⊖ 17 (Enter)

5. $-10 - 17 =$ _____   _____

6. $-10 + (-17) =$ _____   _____

7. $-27 + 220 =$ _____   _____

8. $19 - 43 =$ _____   _____

9. $-35 - (-35) =$ _____   _____

10. $72 + (-47) =$ _____   _____

Addition and Subtraction Using a Calculator (cont.)

Solve. Use your calculator.

11. $3.65 - 2.02 =$ _____

12. $10 - (-5) =$ _____

13. $-901 - 199 =$ _____

14. $-7.1 + 18.6 =$ _____

15. $-2 + (-13) + 7 =$ _____

16. $2 - 7 - (-15) =$ _____

17. $41 / 328 =$ _____

18. $3 * 3.14 =$ _____

19. $-41 / 328 =$ _____

20. $- (3 * 3.14) =$ _____

21. $41 * (7 + 2) =$ _____

22. $41 * (7 + (-2)) =$ _____

Number Stories

23. A salesperson is often assigned a quota. A quota is the dollar value of the goods that the salesperson is expected to sell.

Suppose a salesperson is $3,500 below quota and then makes a sale of $4,700.

Did the salesperson exceed or fall short of his or her quota? _____

Write a number model to figure out by how much the salesperson exceeded or fell short. (Use negative and positive numbers. Think about a number line with the quota at 0.)

Number model: _____

Solution: _____

24. Stock prices change every day. The first day, a stock's price went up $\frac{1}{4}$ dollar per share. The next day, it went down $\frac{1}{2}$ dollar. The third day, it went up $\frac{5}{8}$ dollar.

Did the value increase or decrease from the beginning of Day 1 to the end of

Day 3? _____

Write a number model to figure out by how much the stock increased or decreased over the 3-day period. (Use negative and positive numbers. Think about a number line with the Day 1 starting price at 0.)

Number model: _____

Solution: _____

Use with Lesson 7.10.

Plotting Ordered Pairs

1. Plot the following ordered pairs on the grid below. As you plot each point, connect it with a line segment to the last one you plotted. (Use your ruler.)

   (0,3); (3,−2); (3,−8); (−3,−8); (−3,−2)

2. Plot the following ordered pairs on the grid below. As you plot each point, connect it with a line segment to the last one you plotted. (Use your ruler.)

   $(1,1\frac{1}{3})$; (1,4); (2,4)

3. Plot the following ordered pairs on the grid below. As you plot each point, connect it with a line segment to the last one you plotted. (Use your ruler.)

   (1,−8); (1,−5); (−1,−5); (−1,−8)

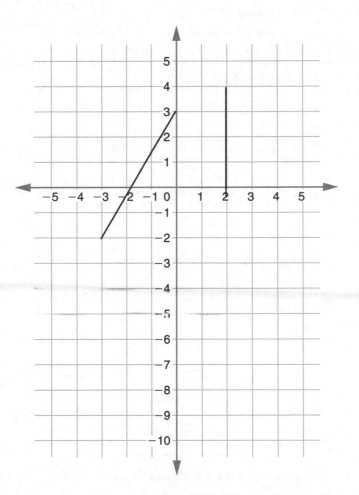

Math Boxes 7.10

1. There are 36 stamps in each package. How many stamps are there in …

   a. $\frac{3}{4}$ of a package? _____

   b. $\frac{5}{6}$ of a package? _____

   c. $\frac{2}{9}$ of a package? _____

   d. $\frac{7}{12}$ of a package? _____

   e. $\frac{2}{3}$ of a package? _____

2. Make true sentences by inserting parentheses.

   a. $19 + 41 * 3 = 180$

   b. $5 = 16 / 2 + 2 - 5$

   c. $-1 = 16 / 2 + 2 - 5$

   d. $24 \div 8 + 4 * 3 = 6$

   e. $24 \div 8 + 4 * 3 = 15$

3. Make a circle graph of the survey results.

Time Spent on Homework	
Time	Percent of Students
0–29 minutes	25
30–59 minutes	48
60–89 minutes	10
90–119 minutes	12
2 hours or more	5

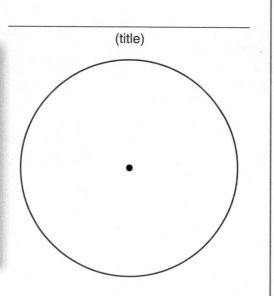

_____
(title)

4. Add. Use fraction sticks to help you.

   a. $\frac{1}{4} + \frac{1}{2} =$ _____

   b. $\frac{1}{4} + \frac{5}{8} =$ _____

   c. $\frac{4}{6} + \frac{1}{3} =$ _____

   d. $\frac{1}{2} + \frac{1}{3} =$ _____

   e. $\frac{1}{6} + \frac{1}{2} =$ _____

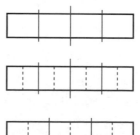

Use with Lesson 7.10.

Time to Reflect

1. Look through your journal pages in this unit. Which pages do you think show your best work? Explain. _____

   _____

   _____

   _____

   _____

   _____

   _____

2. Explain why you think we have negative numbers in our number system. Give examples to support your claims. _____

   _____

   _____

   _____

   _____

   _____

   _____

3. List some of the skills and concepts you learned in this unit that you think are important to remember because you will use them in the future. Explain your answers. _____

   _____

   _____

   _____

   _____

   _____

   _____

Math Boxes 7.11

1. Rewrite each fraction pair with common denominators.

a. $\frac{2}{5}$ and $\frac{3}{7}$ _____

b. $\frac{4}{12}$ and $\frac{6}{9}$ _____

c. $\frac{8}{10}$ and $\frac{10}{15}$ _____

2. **a.** What is the perimeter of the rectangle?

_____

6 units

9 units

b. What is the area? _____

3. Multiply. Use the partial-products algorithm.

a.
$$\begin{array}{r} 26 \\ * \ 32 \\ \hline \end{array}$$

b.
$$\begin{array}{r} 71 \\ * \ 58 \\ \hline \end{array}$$

c.
$$\begin{array}{r} 93 \\ * \ 47 \\ \hline \end{array}$$

4. Write each fraction as a mixed number or a whole number.

a. $\frac{39}{4} = $ _____

b. $\frac{62}{7} = $ _____

c. $\frac{45}{6} = $ _____

d. $\frac{200}{5} = $ _____

e. $\frac{83}{9} = $ _____

5. Find the whole set.

a. 10 is $\frac{1}{5}$ of the set. _____

b. 12 is $\frac{3}{4}$ of the set. _____

c. 8 is $\frac{2}{7}$ of the set. _____

d. 15 is $\frac{5}{8}$ of the set. _____

e. 9 is $\frac{3}{5}$ of the set. _____

Use with Lesson 7.11.

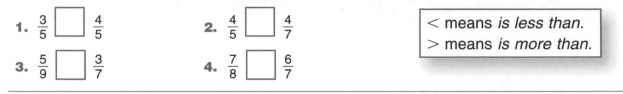

Comparing Fractions

Math Message

Write < or >. Be prepared to explain how you decided on each answer.

1. $\frac{3}{5}$ ☐ $\frac{4}{5}$

2. $\frac{4}{5}$ ☐ $\frac{4}{7}$

3. $\frac{5}{9}$ ☐ $\frac{3}{7}$

4. $\frac{7}{8}$ ☐ $\frac{6}{7}$

> < means *is less than.*
> \> means *is more than.*

Equivalent Fractions

Cross out the fraction in each list that is not equivalent to the other fractions.

5. $\frac{2}{3}$, $\frac{4}{6}$, $\frac{18}{24}$, $\frac{20}{30}$

6. $\frac{1}{4}$, $\frac{2}{8}$, $\frac{4}{20}$, $\frac{6}{24}$, $\frac{8}{32}$

7. $\frac{3}{5}$, $\frac{6}{10}$, $\frac{9}{20}$, $\frac{15}{25}$

Write = or ≠ in each box.

8. $\frac{3}{5}$ ☐ $\frac{10}{15}$

9. $\frac{6}{8}$ ☐ $\frac{16}{24}$

10. $\frac{15}{24}$ ☐ $\frac{5}{8}$

11. $\frac{6}{14}$ ☐ $\frac{2}{7}$

> ≠ means *is not equal to.*

Give three equivalent fractions for each fraction.

12. $\frac{6}{9}$ _____ , _____ , _____

13. $\frac{50}{100}$ _____ , _____ , _____

14. $\frac{7}{10}$ _____ , _____ , _____

15. $\frac{15}{18}$ _____ , _____ , _____

Fill in the missing number.

16. $\frac{3}{4} = \frac{☐}{36}$

17. $\frac{3}{5} = \frac{☐}{20}$

18. $5 = \frac{☐}{2}$

19. $\frac{☐}{9} = \frac{24}{18}$

20. $\frac{9}{12} = \frac{☐}{4}$

21. $\frac{16}{☐} = \frac{8}{10}$

22. $\frac{2}{5} = \frac{6}{☐}$

23. $\frac{15}{☐} = \frac{3}{5}$

24. $\frac{4}{9} = \frac{16}{☐}$

Write < or >.

25. $\frac{2}{5}$ ☐ $\frac{5}{10}$

26. $\frac{3}{4}$ ☐ $\frac{5}{6}$

27. $\frac{3}{8}$ ☐ $\frac{2}{7}$

28. $\frac{3}{5}$ ☐ $\frac{4}{7}$

Fraction Review

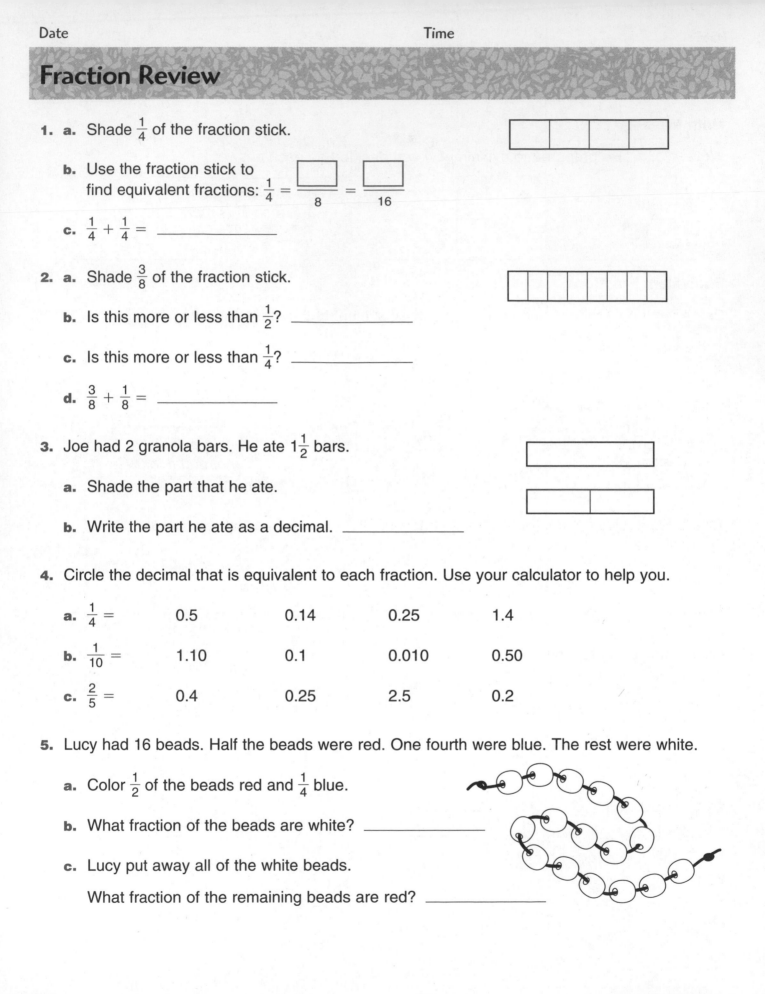

1. a. Shade $\frac{1}{4}$ of the fraction stick.

   b. Use the fraction stick to find equivalent fractions: $\frac{1}{4} = \frac{\boxed{}}{8} = \frac{\boxed{}}{16}$

   c. $\frac{1}{4} + \frac{1}{4} =$ _____

2. a. Shade $\frac{3}{8}$ of the fraction stick.

   b. Is this more or less than $\frac{1}{2}$? _____

   c. Is this more or less than $\frac{1}{4}$? _____

   d. $\frac{3}{8} + \frac{1}{8} =$ _____

3. Joe had 2 granola bars. He ate $1\frac{1}{2}$ bars.

   a. Shade the part that he ate.

   b. Write the part he ate as a decimal. _____

4. Circle the decimal that is equivalent to each fraction. Use your calculator to help you.

   a. $\frac{1}{4} =$ 0.5   0.14   0.25   1.4

   b. $\frac{1}{10} =$ 1.10   0.1   0.010   0.50

   c. $\frac{2}{5} =$ 0.4   0.25   2.5   0.2

5. Lucy had 16 beads. Half the beads were red. One fourth were blue. The rest were white.

   a. Color $\frac{1}{2}$ of the beads red and $\frac{1}{4}$ blue.

   b. What fraction of the beads are white? _____

   c. Lucy put away all of the white beads.

   What fraction of the remaining beads are red? _____

                                       Use with Lesson 8.1.

Math Boxes 8.1

1. Use a straightedge to draw as many lines of symmetry as you can.

SRB
149

2. Round each number to the nearest hundredth.

a. 432.089 _____

b. 650.127 _____

c. 227.715 _____

d. 38.002 _____

e. 61.099 _____

SRB
45 46

3. Solve only the problems with an answer over 1,000.

a.
```
  429
+ 813
```

b.
```
  729
+ 202
```

c.
```
  914
+ 986
```

d.
```
  1,235
-   189
```

e.
```
  1,605
-   493
```

SRB
13–17
226 227

4. Use the grid on the right to locate the following objects on the map. The first one has been done for you.

a. Fifth grader _D4_

b. Boat _____

c. Car _____

d. House _____

e. Tree _____

Addition of Fractions

Math Message

Add. Write the sums in simplest form.

1. $\frac{3}{5} + \frac{1}{5} =$ _____

2. $\frac{3}{8} + \frac{1}{8} =$ _____

3. $\frac{2}{3} + \frac{2}{3} + \frac{2}{3} =$ _____

4. $\frac{3}{7} + \frac{5}{7} =$ _____

5. $\frac{7}{10} + \frac{7}{10} =$ _____

6. $\frac{5}{9} + \frac{7}{9} =$ _____

7. $\frac{1}{6} + \frac{2}{3} =$ _____

8. $\frac{2}{3} + \frac{2}{5} =$ _____

9. $\frac{5}{6} + \frac{5}{8} =$ _____

Addition of Mixed Numbers

Add. Write each sum as a whole number or mixed number.

10. $1\frac{3}{5}$
   $+ 1\frac{1}{5}$

11. $1\frac{1}{2}$
   $+ \frac{1}{2}$

12. $2\frac{1}{4}$
   $+ 3\frac{3}{4}$

Fill in the missing numbers.

13. $5\frac{12}{7} = 6\,\dfrac{5}{\boxed{}}$

14. $7\frac{8}{5} = \boxed{}\,\frac{3}{5}$

15. $2\frac{5}{4} = 3\,\dfrac{\boxed{}}{4}$

16. $4\frac{5}{3} = 5\,\dfrac{\boxed{}}{3}$

17. $12\frac{11}{6} = 13\,\dfrac{\boxed{}}{6}$

18. $9\frac{13}{10} = 10\,\dfrac{\boxed{}}{10}$

Add. Write each sum as a mixed number in simplest form.

19. $3\frac{2}{3}$
   $+ 5\frac{2}{3}$

20. $4\frac{6}{7}$
   $+ 2\frac{4}{7}$

21. $3\frac{4}{9}$
   $+ 6\frac{8}{9}$

Use with Lesson 8.2.

Addition of Mixed Numbers (cont.)

To add mixed numbers in which the fractions do not have the same denominator, you must first rename one or both fractions so that both fractions have a common denominator.

Example $2\frac{3}{5} + 4\frac{2}{3} = ?$

- Find a common denominator: The QCD of $\frac{3}{5}$ and $\frac{2}{3}$ is $5 * 3 = 15$.

- Write the problem in vertical form and rename the fractions:

$$
\begin{array}{r} 2\frac{3}{5} \\ + \ 4\frac{2}{3} \\ \hline \end{array}
\quad \rightarrow \quad
\begin{array}{r} 2\frac{9}{15} \\ + \ 4\frac{10}{15} \\ \hline 6\frac{19}{15} \end{array}
$$

- Add.

- Rename the sum. $6\frac{19}{15} = 7\frac{4}{15}$

Add. Write each sum as a mixed number in simplest form. Show your work.

1. $2\frac{1}{3} + 3\frac{1}{4} = $ _____

2. $5\frac{1}{2} + 2\frac{2}{5} = $ _____

3. $6\frac{1}{3} + 2\frac{4}{9} = $ _____

4. $1\frac{1}{2} + 4\frac{3}{4} = $ _____

5. $7\frac{1}{4} + 2\frac{5}{6} = $ _____

6. $3\frac{5}{6} + 3\frac{3}{4} = $ _____

Reading a Ruler

On the ruler below, points *A* through *L* mark distances from the beginning of the ruler (0 inches). Give the distance from 0 for each point. Point *A* has been done for you.

1. *A:* $4\frac{1}{4}$ in.

2. *B:* _____ in.

3. *C:* _____ in.

4. *D:* _____ in.

5. *E:* _____ in.

6. *F:* _____ in.

7. *G:* _____ in.

8. *H:* _____ in.

9. *I:* _____ in.

10. *J:* _____ in.

11. *K:* _____ in.

12. *L:* _____ in.

Pick four of the points in Problems 1–12. For each point, write an equivalent name for its distance from 0.

Example   *A* : $4\frac{1}{4}$ in. = $4\frac{4}{16}$ in.

13. ____ : _____ in. = _____ in.

14. ____ : _____ in. = _____ in.

15. ____ : _____ in. = _____ in.

16. ____ : _____ in. = _____ in.

Write >, <, or =.

17. $\frac{3}{4}$ _____ $\frac{7}{8}$

18. $\frac{6}{9}$ _____ $\frac{8}{12}$

19. $\frac{4}{7}$ _____ $\frac{5}{9}$

20. $\frac{13}{24}$ _____ $\frac{7}{15}$

21. $\frac{5}{7}$ _____ $\frac{7}{10}$

22. $\frac{17}{18}$ _____ $\frac{9}{10}$

23. Explain how you got your answer for Problem 20. _____

_____

_____

Math Boxes 8.2

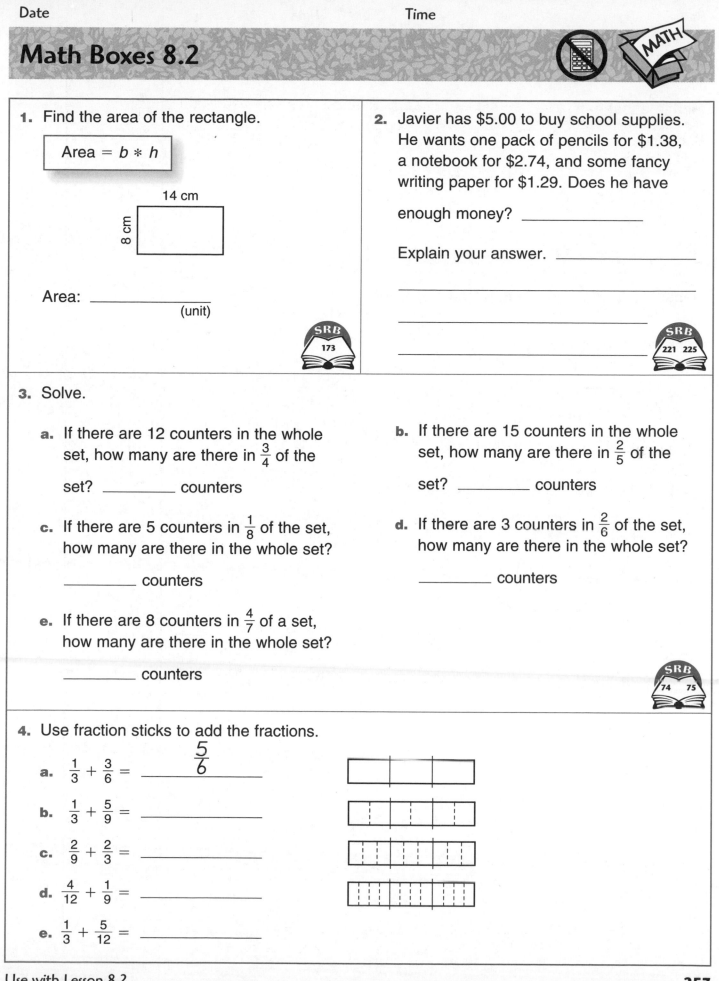

1. Find the area of the rectangle.

   Area = b * h

   14 cm

   8 cm

   Area: _____
              (unit)

   SRB 173

2. Javier has $5.00 to buy school supplies. He wants one pack of pencils for $1.38, a notebook for $2.74, and some fancy writing paper for $1.29. Does he have

   enough money? _____

   Explain your answer. _____

   _____

   _____

   _____

   SRB 221 225

3. Solve.

   a. If there are 12 counters in the whole set, how many are there in $\frac{3}{4}$ of the

      set? _____ counters

   b. If there are 15 counters in the whole set, how many are there in $\frac{2}{5}$ of the

      set? _____ counters

   c. If there are 5 counters in $\frac{1}{8}$ of the set, how many are there in the whole set?

      _____ counters

   d. If there are 3 counters in $\frac{2}{6}$ of the set, how many are there in the whole set?

      _____ counters

   e. If there are 8 counters in $\frac{4}{7}$ of a set, how many are there in the whole set?

      _____ counters

   SRB 74 75

4. Use fraction sticks to add the fractions.

   a. $\frac{1}{3} + \frac{3}{6} = $ _____$\frac{5}{6}$_____

   b. $\frac{1}{3} + \frac{5}{9} = $ _____

   c. $\frac{2}{9} + \frac{2}{3} = $ _____

   d. $\frac{4}{12} + \frac{1}{9} = $ _____

   e. $\frac{1}{3} + \frac{5}{12} = $ _____

Subtraction of Mixed Numbers

Math Message

Subtract.

1. $3\frac{3}{4}$
    $-1\frac{1}{4}$
    _____

2. $4\frac{4}{5}$
    -2
    _____

3. $7\frac{5}{6}$
    $-2\frac{2}{6}$
    _____

Renaming and Subtracting Mixed Numbers

Fill in the missing numbers.

4. $5\frac{1}{4} = 4\,\dfrac{\boxed{}}{4}$

5. $6 = 5\,\dfrac{\boxed{}}{3}$

6. $3\frac{5}{6} = \dfrac{\boxed{}}{6}$

7. $8\frac{7}{9} = \boxed{}\,\dfrac{16}{9}$

Subtract. Write your answers in simplest form. Show your work.

8. $8 - \frac{1}{3} =$ _____

9. $5 - 2\frac{3}{5} =$ _____

10. $7\frac{1}{4} - 3\frac{3}{4} =$ _____

11. $4\frac{5}{8} - 3\frac{7}{8} =$ _____

12. $6\frac{2}{9} - 4\frac{5}{9} =$ _____

13. $10\frac{3}{10} - 5\frac{7}{10} =$ _____

Use with Lesson 8.3.

Addition and Subtraction Patterns

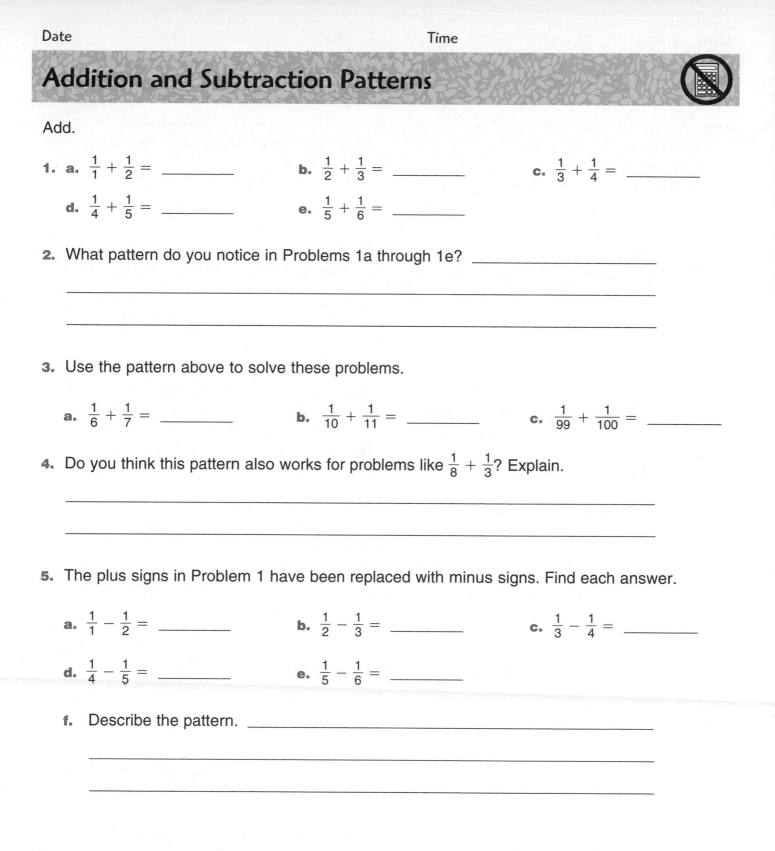

Add.

1. **a.** $\frac{1}{1} + \frac{1}{2} =$ _____   **b.** $\frac{1}{2} + \frac{1}{3} =$ _____   **c.** $\frac{1}{3} + \frac{1}{4} =$ _____

     d. $\frac{1}{4} + \frac{1}{5} =$ _____   **e.** $\frac{1}{5} + \frac{1}{6} =$ _____

2. What pattern do you notice in Problems 1a through 1e? _____

_____

_____

3. Use the pattern above to solve these problems.

     a. $\frac{1}{6} + \frac{1}{7} =$ _____   **b.** $\frac{1}{10} + \frac{1}{11} =$ _____   **c.** $\frac{1}{99} + \frac{1}{100} =$ _____

4. Do you think this pattern also works for problems like $\frac{1}{8} + \frac{1}{3}$? Explain.

_____

_____

5. The plus signs in Problem 1 have been replaced with minus signs. Find each answer.

     a. $\frac{1}{1} - \frac{1}{2} =$ _____   **b.** $\frac{1}{2} - \frac{1}{3} =$ _____   **c.** $\frac{1}{3} - \frac{1}{4} =$ _____

     d. $\frac{1}{4} - \frac{1}{5} =$ _____   **e.** $\frac{1}{5} - \frac{1}{6} =$ _____

     f. Describe the pattern. _____

_____

_____

Math Boxes 8.3

1. Use a straightedge to draw as many lines of symmetry as you can.

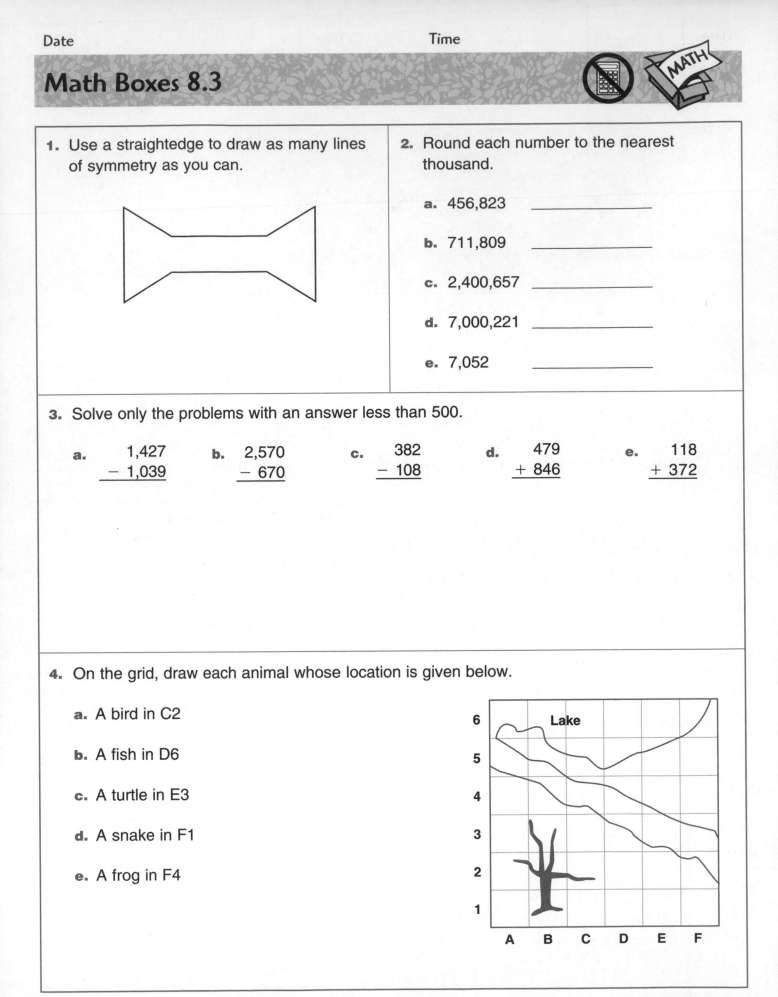

2. Round each number to the nearest thousand.

a. 456,823 _____

b. 711,809 _____

c. 2,400,657 _____

d. 7,000,221 _____

e. 7,052 _____

3. Solve only the problems with an answer less than 500.

a.
$$\begin{array}{r} 1,427 \\ -\ 1,039 \\ \hline \end{array}$$

b.
$$\begin{array}{r} 2,570 \\ -\ 670 \\ \hline \end{array}$$

c.
$$\begin{array}{r} 382 \\ -\ 108 \\ \hline \end{array}$$

d.
$$\begin{array}{r} 479 \\ +\ 846 \\ \hline \end{array}$$

e.
$$\begin{array}{r} 118 \\ +\ 372 \\ \hline \end{array}$$

4. On the grid, draw each animal whose location is given below.

a. A bird in C2

b. A fish in D6

c. A turtle in E3

d. A snake in F1

e. A frog in F4

Use with Lesson 8.3.

Calculator Key Investigation

Explore the seven function keys on your calculator. Use the sample keystrokes to help you find what each key does.

Key(s)	Sample Keystrokes	Function of Key(s)
(Unit)	2 (Unit) 1 (n) 2 (d) (Enter)	
(n) and (d)	2 (n) 3 (d) (Enter)	
(F↔D)	4 (n) 5 (d) (Enter) (F↔D) (F↔D)	
(Simp)	4 (n) 12 (d) (Simp) (Enter) (Simp) (Enter)	
(U$\frac{n}{d}$↔$\frac{n}{d}$)	14 (n) 3 (d) (Enter) (Simp) (Enter) (U$\frac{n}{d}$↔$\frac{n}{d}$) (U$\frac{n}{d}$↔$\frac{n}{d}$)	
(Fac)	4 (n) 8 (Simp) (Enter) (Fac)	

Finding a Fraction of a Number

One way to find a fraction of a number is to use a **unit fraction.** (A unit fraction is a fraction with 1 in the numerator.) You can also use a diagram to help you understand the problem.

Example What is $\frac{7}{8}$ of 32?

$\frac{1}{8}$ of 32 is 4. So $\frac{7}{8}$ of 32 is 7 * 4 = 28.

32

?

Solve.

1. $\frac{1}{5}$ of 75 = _____

2. $\frac{2}{5}$ of 75 = _____

3. $\frac{4}{5}$ of 75 = _____

4. $\frac{1}{8}$ of 120 = _____

5. $\frac{3}{8}$ of 120 = _____

6. $\frac{5}{8}$ of 120 = _____

Solve Problems 7–18. They come from a math book that was published in 1904.

• First think of $\frac{1}{3}$ of each of these numbers, and then state $\frac{2}{3}$ of each.

7. 9 _____

8. 6 _____

9. 12 _____

10. 3 _____

11. 21 _____

12. 30 _____

• First think of $\frac{1}{4}$ of each of these numbers, and then state $\frac{3}{4}$ of each.

13. 32 _____

14. 40 _____

15. 12 _____

16. 24 _____

17. 20 _____

18. 28 _____

19. Lydia has 7 pages of a 12-page song memorized. Has she memorized more than $\frac{2}{3}$ of the song? _____

20. A CD that normally sells for $15 is on sale for $\frac{1}{3}$ off. What is the sale price? _____

21. Christine bought a coat for $\frac{1}{4}$ off the regular price. She saved $20. What did she pay for the coat? _____

Use with Lesson 8.4.

Math Boxes 8.4

1. Find the area of the rectangle.

$$Area = b * h$$

12 m

6 m

Area: _____
            (unit)

2. Julie makes $4.00 per week for doing the dishes every night. She paid her sister Amy $0.75 each time Amy did the dishes for her. Is that a fair price?

_____

Explain your answer. _____

_____

_____

_____

3. Solve.

a. If there are 18 counters in the whole set, how many are there in $\frac{5}{6}$ of the set? _____ counters

b. If there are 21 counters in the whole set, how many are there in $\frac{2}{3}$ of the set? _____ counters

c. If there are 6 counters in $\frac{2}{7}$ of the set, how many are there in the whole set?

_____ counters

d. If there are 10 counters in $\frac{1}{5}$ of the set, how many are there in the whole set?

_____ counters

e. If there are 9 counters in $\frac{3}{8}$ of the set, how many are there in the whole set?

_____ counters

4. Use fraction sticks to add the fractions.

a. $\frac{1}{4} + \frac{3}{3} =$ _____

b. $\frac{1}{8} + \frac{1}{2} =$ _____

c. $\frac{5}{8} + \frac{1}{4} =$ _____

d. $\frac{5}{12} + \frac{1}{4} =$ _____

e. $\frac{3}{4} + \frac{1}{12} =$ _____

Math Boxes 8.5

1. Name 2 objects that are shaped like a rectangular prism.

_____

_____

_____

2. Divide mentally.

a.   382 / 7 → _____

b.   795 / 5 → _____

c.   496 / 4 → _____

d.   283 ÷ 6 → _____

e.   1,625 ÷ 8 → _____

SRB
140

3. Amanda found a can containing 237 dominoes. A full set has 28 dominoes. What is the

greatest number of complete sets that can be in the can? _____

(unit)

Explain how you found your answer. _____

_____

_____

_____

SRB
22–24
224

4. Complete the table.

Standard Notation	Exponential Notation
10,000	
	10^3
	10^8
1,000,000,000	
	10^5

SRB
5

5. Complete the "What's My Rule?" table and state the rule.

Rule: _____

in	out
3	
8	40
$\frac{1}{2}$	
	50
4	

SRB
215 216

Use with Lesson 8.5.

Number-Line Models

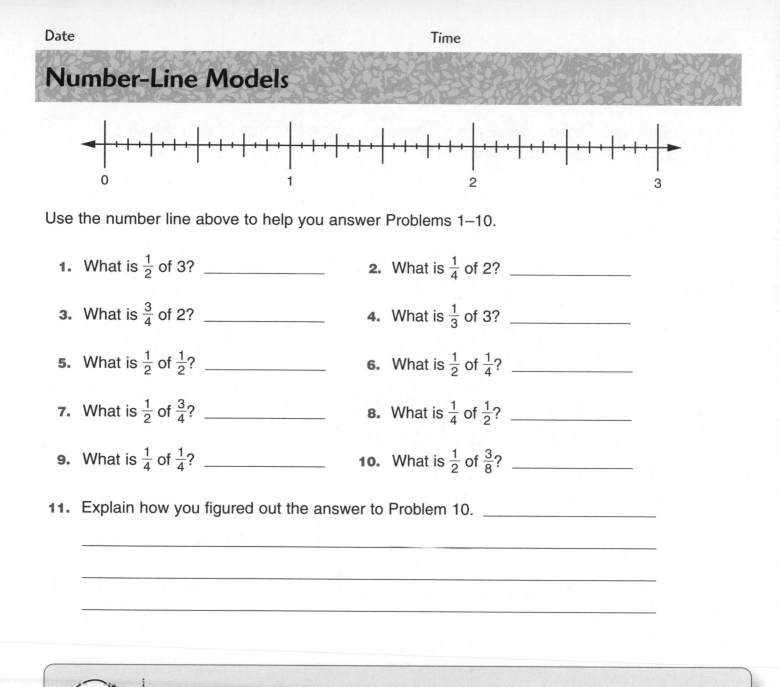

Use the number line above to help you answer Problems 1–10.

1. What is $\frac{1}{2}$ of 3? _____

2. What is $\frac{1}{4}$ of 2? _____

3. What is $\frac{3}{4}$ of 2? _____

4. What is $\frac{1}{3}$ of 3? _____

5. What is $\frac{1}{2}$ of $\frac{1}{2}$? _____

6. What is $\frac{1}{2}$ of $\frac{1}{4}$? _____

7. What is $\frac{1}{2}$ of $\frac{3}{4}$? _____

8. What is $\frac{1}{4}$ of $\frac{1}{2}$? _____

9. What is $\frac{1}{4}$ of $\frac{1}{4}$? _____

10. What is $\frac{1}{2}$ of $\frac{3}{8}$? _____

11. Explain how you figured out the answer to Problem 10. _____

_____

_____

_____

All (Winged) Creatures: Great and Small

The smallest bird is the bee hummingbird. It weighs about 2 grams (0.07 ounces) and is about 5.5 centimeters (2.2 inches) long. The largest bird is the ostrich. An adult ostrich can weigh more than 150 kilograms (330 pounds) and stand 2.5 meters (8 feet) tall. (It's no wonder that ostriches can't fly!) It would take more than 75,000 bee hummingbirds to balance one ostrich (if you could find a balance big enough).

Source: Britannica Online

Paper-Folding Problems

Record your work for the four fraction problems you solved by paper folding. Sketch the folds and shading. Write an X on the parts that show the answer.

1. $\frac{1}{2}$ of $\frac{1}{2}$ is _____.

2. $\frac{2}{3}$ of $\frac{1}{2}$ is _____.

3. $\frac{1}{4}$ of $\frac{2}{3}$ is _____.

4. $\frac{3}{4}$ of $\frac{1}{2}$ is _____.

Use with Lesson 8.5.

Paper-Folding Problems (cont.)

Solve these problems by paper folding. Sketch the folds and shading. Write an X on the parts that show the answer.

5. $\frac{1}{3}$ of $\frac{3}{4}$ is _____.

6. $\frac{1}{8}$ of $\frac{1}{2}$ is _____.

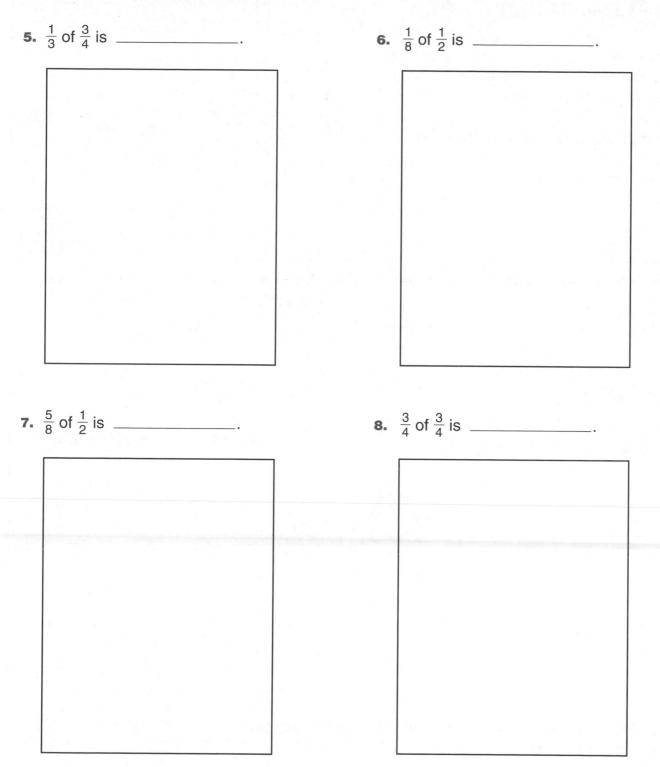

7. $\frac{5}{8}$ of $\frac{1}{2}$ is _____.

8. $\frac{3}{4}$ of $\frac{3}{4}$ is _____.

Use with Lesson 8.5.

Math Boxes 8.6

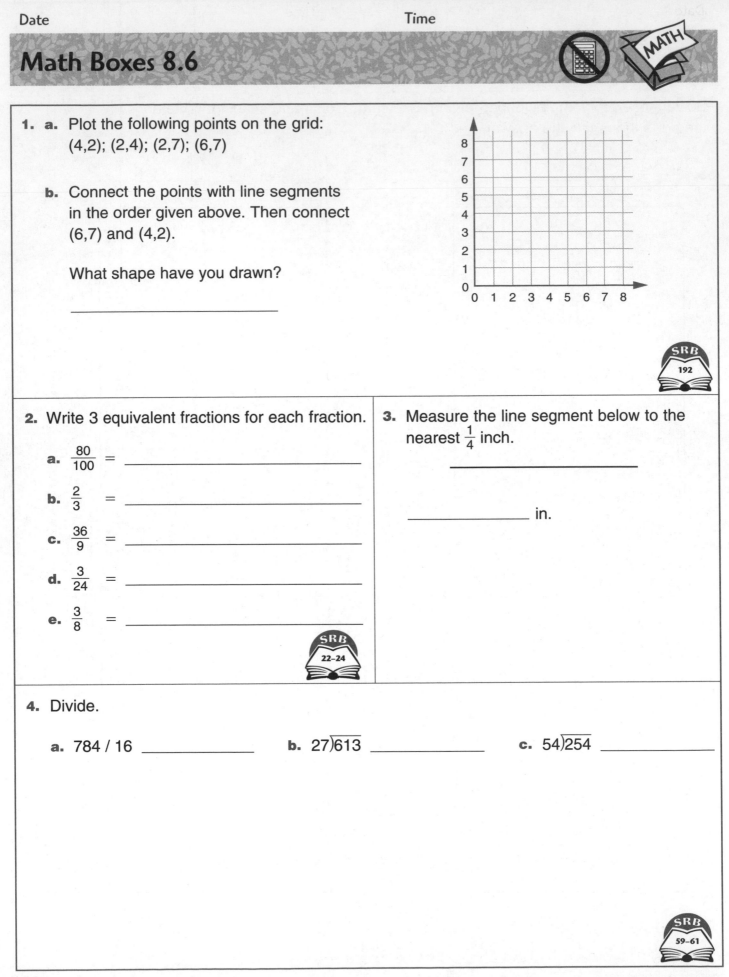

1. **a.** Plot the following points on the grid:
(4,2); (2,4); (2,7); (6,7)

   b. Connect the points with line segments in the order given above. Then connect (6,7) and (4,2).

   What shape have you drawn?

   _____

   SRB
   192

2. Write 3 equivalent fractions for each fraction.

   a. $\frac{80}{100}$ = _____

   b. $\frac{2}{3}$ = _____

   c. $\frac{36}{9}$ = _____

   d. $\frac{3}{24}$ = _____

   e. $\frac{3}{8}$ = _____

   SRB
   22–24

3. Measure the line segment below to the nearest $\frac{1}{4}$ inch.

   _____

   _____ in.

4. Divide.

   a. 784 / 16 _____

   b. 27$\overline{)613}$ _____

   c. 54$\overline{)254}$ _____

   SRB
   59–61

Use with Lesson 8.6.

Fraction Multiplication

1. Use the rectangle at the right to sketch how you would fold paper to help you find $\frac{1}{3}$ of $\frac{2}{3}$.

   What is $\frac{1}{3}$ of $\frac{2}{3}$? _____

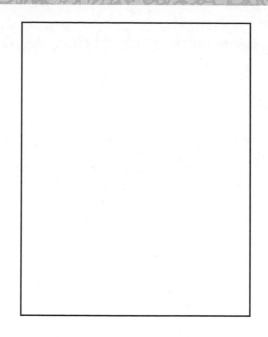

2. Use the rectangle at the right to sketch how you would fold paper to help you find $\frac{1}{4}$ of $\frac{3}{5}$.

   What is $\frac{1}{4}$ of $\frac{3}{5}$? _____

3. Rewrite "$\frac{2}{3}$ of $\frac{3}{4}$" using the multiplication symbol *. _____

4. Rewrite the following using the multiplication symbol *.

   a. $\frac{1}{4}$ of $\frac{1}{3}$ _____   b. $\frac{4}{5}$ of $\frac{2}{3}$ _____

   c. $\frac{1}{6}$ of $\frac{1}{4}$ _____   d. $\frac{3}{7}$ of $\frac{2}{5}$ _____

An Area Model for Fraction Multiplication

1. Use the rectangle at the right to find $\frac{2}{3} * \frac{3}{4}$.

$\frac{2}{3} * \frac{3}{4} =$ _____

Your completed drawing in Problem 1 is called an **area model.**

Use area models to complete the following.

2.

$\frac{2}{3} * \frac{1}{5} =$ _____

3.

$\frac{3}{4} * \frac{2}{5} =$ _____

4.

$\frac{1}{4} * \frac{5}{6} =$ _____

5.

$\frac{3}{8} * \frac{3}{5} =$ _____

6.

$\frac{1}{2} * \frac{5}{8} =$ _____

7.

$\frac{5}{6} * \frac{4}{5} =$ _____

8. Make up your own fraction multiplication problem.
Use an area model to help you solve it.

_____

Use with Lesson 8.6.

An Algorithm for Fraction Multiplication

1. Look carefully at the fractions on journal page 270. What is the relationship between the numerators and the denominators of the two fractions being multiplied and the numerator and the denominator of their product?

   _____

   _____

   _____

   _____

2. Describe a way to multiply two fractions. _____

   _____

   _____

   _____

3. Multiply the following fractions, using the shortcut discussed in class.

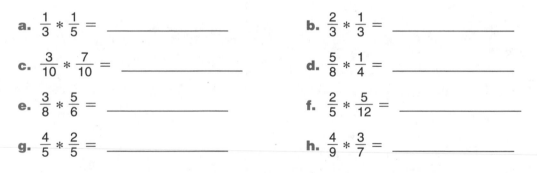

   a. $\dfrac{1}{3} * \dfrac{1}{5} =$ _____

   b. $\dfrac{2}{3} * \dfrac{1}{3} =$ _____

   c. $\dfrac{3}{10} * \dfrac{7}{10} =$ _____

   d. $\dfrac{5}{8} * \dfrac{1}{4} =$ _____

   e. $\dfrac{3}{8} * \dfrac{5}{6} =$ _____

   f. $\dfrac{2}{5} * \dfrac{5}{12} =$ _____

   g. $\dfrac{4}{5} * \dfrac{2}{5} =$ _____

   h. $\dfrac{4}{9} * \dfrac{3}{7} =$ _____

Cashing in on Fractions

Torn money might be worth something. If you have $\dfrac{3}{5}$ or more of a bill, it can be redeemed for its full face value. If you have less than $\dfrac{3}{5}$ but more than $\dfrac{2}{5}$ of a bill, it can be redeemed for $\dfrac{1}{2}$ of its face value. If you have $\dfrac{2}{5}$ or less, your bill is worthless.

Source: You Can't Count a Billion Dollars

Place-Value Practice

Find the missing number.

1. The digit in the tens place is twice as big as the digit in the tenths place.
The digit in the ones place is $\frac{1}{2}$ the digit in the tenths place.
The digit in the hundredths place is $\frac{1}{2}$ the digit in the ones place.
The digit in the hundreds place is the largest odd digit.

___ ___ ___ . ___ ___

2. The digit in the hundreds place is a square number and it is odd.
The digit in the tens place is 1 more than the square root of 16.
The digit in the hundredths place is 0.1 larger than $\frac{1}{10}$ of the digit in the
hundreds place.
The digit in the thousandths place is equivalent to $\frac{30}{5}$.
The other digits are all 2s.

___ ___ ___ . ___ ___ ___

3. Record the calculator keystrokes you enter to make the changes described below.

Beginning Number	Change to	Keystrokes
12,204	15,204	12204
807,995	808,005	807995
2.112	2.712	2.112
17.054	18.104	17.054
34.921	35.021	34.921

Use with Lesson 8.6.

A Blast from the Past

1. From *Kindergarten Everyday Mathematics:*

   This slice of pizza is what
   fraction of the whole pizza? _____

   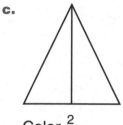

2. From *First Grade Everyday Mathematics:*

   Write a fraction in each part of the diagrams below. Then color the
   figures as directed.

   a.

   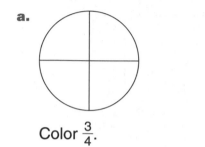

   Color $\frac{3}{4}$.

   b.

   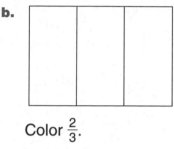

   Color $\frac{2}{3}$.

   c.

   Color $\frac{2}{2}$.

3. From *Second Grade Everyday Mathematics:*

   a.

   b.

   Color $\frac{1}{4}$ of the beads.   Color $\frac{1}{8}$ of the beads.

4. From *Third Grade Everyday Mathematics:*

   a. $\frac{1}{2}$ of $\frac{1}{4}$ = _____   **b.** $\frac{1}{8}$ of $\frac{1}{2}$ = _____   **c.** $\frac{1}{2}$ of $\frac{1}{8}$ = _____

5. From *Fourth Grade Everyday Mathematics:*

   Cross out $\frac{5}{6}$ of the dimes.

Area Models

Draw an area model for each product. Then write the product as a fraction or as a mixed number.

Example $\frac{2}{3} * 2 =$ _$\frac{4}{3}$, or $1\frac{1}{3}$_

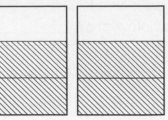

1. $\frac{1}{3} * 4 =$ _____

2. $\frac{1}{4} * 3 =$ _____

3. $2 * \frac{3}{5} =$ _____

4. $\frac{3}{8} * 3 =$ _____

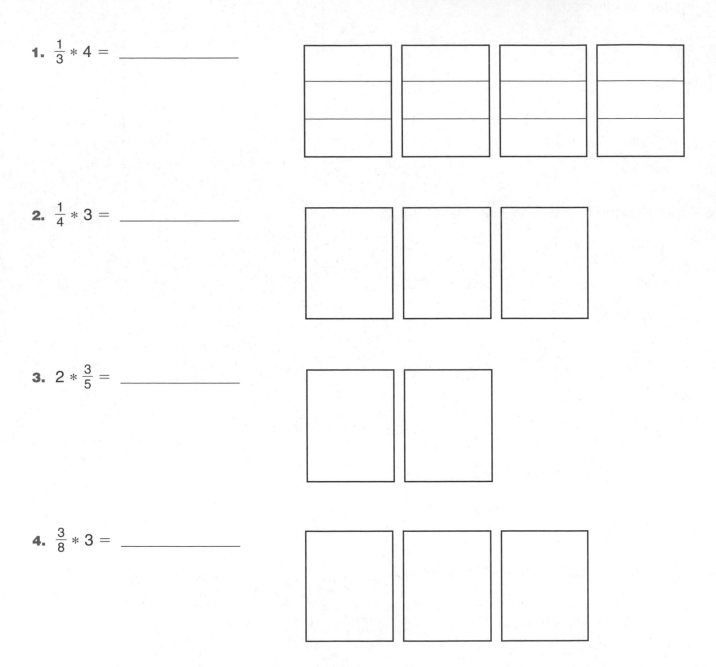

          Use with Lesson 8.7.

Using the Fraction Multiplication Algorithm

An Algorithm for Fraction Multiplication

$$\frac{a}{b} * \frac{c}{d} = \frac{a * c}{b * d}$$

The denominator of the product is the product of the denominators, and the numerator of the product is the product of the numerators.

Example $\frac{2}{3} * 2$

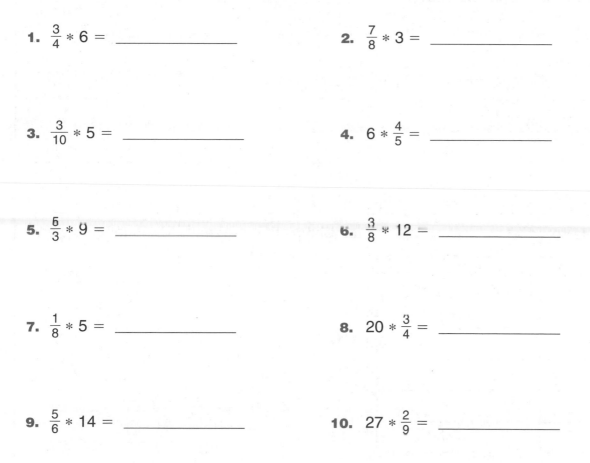

$$\frac{2}{3} * 2 \;\; = \frac{2}{3} * \frac{2}{1} \qquad \text{Think of 2 as } \frac{2}{1}.$$

$$= \frac{2 * 2}{3 * 1} \qquad \text{Apply the algorithm.}$$

$$= \frac{4}{3}, \text{ or } 1\frac{1}{3} \qquad \text{Calculate the numerator and denominator.}$$

Use the fraction multiplication algorithm to calculate the following products.

1. $\frac{3}{4} * 6 =$ _____

2. $\frac{7}{8} * 3 =$ _____

3. $\frac{3}{10} * 5 =$ _____

4. $6 * \frac{4}{5} =$ _____

5. $\frac{5}{3} * 9 =$ _____

6. $\frac{3}{8} * 12 =$ _____

7. $\frac{1}{8} * 5 =$ _____

8. $20 * \frac{3}{4} =$ _____

9. $\frac{5}{6} * 14 =$ _____

10. $27 * \frac{2}{9} =$ _____

Math Boxes 8.7

1. Name 2 objects that are shaped like prisms that are not rectangular prisms.

_____

_____

_____

2. Divide mentally.

a. $472 \div 5 \rightarrow$ _____

b. $384 / 6 \rightarrow$ _____

c. $729 / 8 \rightarrow$ _____

d. $543 \div 4 \rightarrow$ _____

e. $576 \div 9 \rightarrow$ _____

3. Raphael bought 14 pounds of meat to make hamburgers at the Fourth of July barbecue. He made 5 hamburgers from each pound. Buns come in packages of 8. How many

packages of buns did Raphael need? _____

(unit)

Explain your answer. _____

_____

_____

_____

4. Complete the table.

Standard Notation	Exponential Notation
	10^5
	10^9
1,000,000	
10,000	
	10^7

5. Complete the "What's My Rule?" table and state the rule.

Rule: _____

in	out
48	
40	5
	$\frac{1}{8}$
	0
16	

Use with Lesson 8.7.

Date _____ Time _____

You know that fractions larger than 1 can be written in several ways.

Whole
hexagon

Example

If a ⬡ is worth 1,

what is  worth?

The mixed-number name is $3\frac{5}{6}$ ($3\frac{5}{6}$ means $3 + \frac{5}{6}$).

The fraction name is $\frac{23}{6}$. Think sixths:

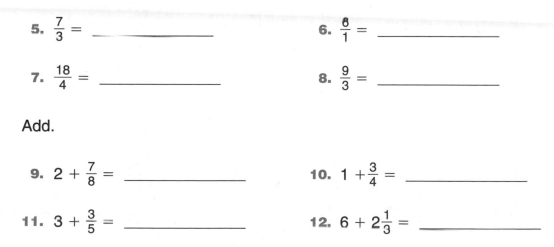

$3\frac{5}{6}$, $3 + \frac{5}{6}$, and $\frac{23}{6}$ are different names for the same number.

Write the following mixed numbers as fractions.

1. $2\frac{3}{5} =$ _____

2. $4\frac{7}{8} =$ _____

3. $1\frac{2}{3} =$ _____

4. $3\frac{6}{4} =$ _____

Write the following fractions as mixed or whole numbers.

5. $\frac{7}{3} =$ _____

6. $\frac{6}{1} =$ _____

7. $\frac{18}{4} =$ _____

8. $\frac{9}{3} =$ _____

Add.

9. $2 + \frac{7}{8} =$ _____

10. $1 + \frac{3}{4} =$ _____

11. $3 + \frac{3}{5} =$ _____

12. $6 + 2\frac{1}{3} =$ _____

Use with Lesson 8.8.

Multiplication of Fractions and Mixed Numbers

Examples Using Partial Products

$2\frac{1}{3} * 2\frac{1}{2} = (2 + \frac{1}{3}) * (2 + \frac{1}{2})$

$\quad 2 * 2 = 4$

$\quad 2 * \frac{1}{2} = 1$

$\quad \frac{1}{3} * 2 = \frac{2}{3}$

$\quad \frac{1}{3} * \frac{1}{2} = + \frac{1}{6}$

$\qquad\qquad\quad 5\frac{5}{6}$

$3\frac{1}{4} * \frac{2}{5} = (3 + \frac{1}{4}) * \frac{2}{5}$

$\quad 3 * \frac{2}{5} = \frac{6}{5} = \quad 1\frac{1}{5}$

$\quad \frac{1}{4} * \frac{2}{5} = \frac{2}{20} = \quad + \frac{1}{10}$

$\qquad\qquad\qquad\qquad 1\frac{3}{10}$

Examples Converting Mixed Numbers to Fractions

$2\frac{1}{3} * 2\frac{1}{2} = \frac{7}{3} * \frac{5}{2}$

$\qquad\qquad = \frac{35}{6} = 5\frac{5}{6}$

$3\frac{1}{4} * \frac{2}{5} = \frac{13}{4} * \frac{2}{5}$

$\qquad\qquad = \frac{26}{20} = 1\frac{6}{20} = 1\frac{3}{10}$

Solve the following fraction and mixed-number multiplication problems.

1. $3\frac{1}{2} * 2\frac{1}{5} =$ _____

2. $10\frac{3}{4} * \frac{1}{2} =$ _____

3. The surface of a calculator is approximately a rectangular prism. The back face has an area of about

_____ in.²

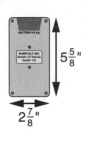

$5\frac{5}{8}"$

$2\frac{7}{8}"$

4. The area of a sheet of notebook paper is about

_____ in.²

$10\frac{1}{2}"$

$8"$

5. The area of a computer disk is about

_____ in.²

$3\frac{5}{6}"$

$3\frac{1}{2}"$

6. The area of the flag is about

_____ yd²

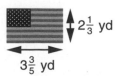

$2\frac{1}{3}$ yd

$3\frac{3}{5}$ yd

7. Is the area of the flag greater or less than the area of your desk or tabletop? _____

                    Use with Lesson 8.8.

Track Records on the Moon and the Planets

Every moon and planet in our solar system pulls objects toward it with the force called **gravity.**

In a recent Olympic games, the winning high jump was 7 feet 8 inches, or $7\frac{2}{3}$ feet. The winning pole vault was 19 feet. Suppose that the Olympics were held on Earth's Moon, or on Jupiter, Mars, or Venus. What height might we expect for a winning high jump or a winning pole vault?

1. On the Moon, one could jump about 6 times as high as on Earth.
   What would be the height of the winning …

   high jump? About _____ feet   pole vault? About _____ feet

2. On Jupiter, one could jump about $\frac{3}{8}$ as high as on Earth.
   What would be the height of the winning …

   high jump? About _____ feet   pole vault? About _____ feet

3. On Mars, one could jump about $2\frac{2}{3}$ times as high as on Earth.
   What would be the height of the winning …

   high jump? About _____ feet   pole vault? About _____ feet

4. On Venus, one could jump about $1\frac{1}{7}$ times as high as on Earth.
   What would be the height of the winning …

   high jump? About _____ feet   pole vault? About _____ feet

5. Is Jupiter's pull of gravity stronger or weaker than Earth's? _____
   Explain your reasoning.

   _____

   _____

   _____

Challenge

6. The winning pole-vault height given above was rounded to the nearest whole number. The actual winning height was 19 feet $\frac{1}{4}$ inch. If you used this actual measurement, about how high would the winning jump be …

   on the Moon? _____   on Jupiter? _____

   on Mars? _____   on Venus? _____

Use with Lesson 8.8.

1. a. Plot the following points on the grid:

   (–3,–3); (1,1); (4,1); (0,–3)

   b. Connect the points with line segments in the order given above. Then connect (–3,–3) and (0,–3).

   What shape have you drawn?

   _____

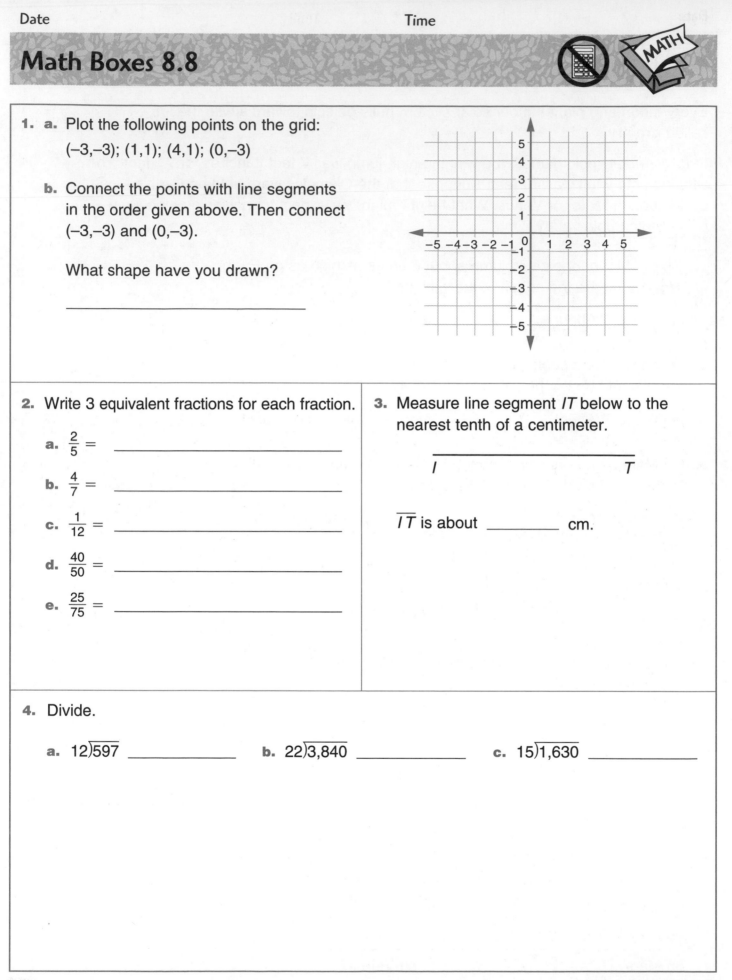

2. Write 3 equivalent fractions for each fraction.

   a. $\frac{2}{5}$ = _____

   b. $\frac{4}{7}$ = _____

   c. $\frac{1}{12}$ = _____

   d. $\frac{40}{50}$ = _____

   e. $\frac{25}{75}$ = _____

3. Measure line segment *IT* below to the nearest tenth of a centimeter.

   I _____ T

   $\overline{IT}$ is about _____ cm.

4. Divide.

   a. $12\overline{)597}$ _____
   b. $22\overline{)3,840}$ _____
   c. $15\overline{)1,630}$ _____

Finding a Percent of a Number

1. The Madison Middle School boys' basketball team has played 5 games. The table at the right shows the number of shots taken by each player and the percent of shots that were baskets. Study the example. Then calculate the number of baskets made by each player.

Example

Bill took 15 shots.
He made a basket on 40% of these shots.

$40\% = \frac{40}{100}$, or $\frac{4}{10}$

$\frac{4}{10}$ of 15 $= \frac{4}{10} * \frac{15}{1} = \frac{4 * 15}{10 * 1} = \frac{60}{10} = 6$

Bill made 6 baskets.

Player	Shots Taken	Percent Made	Baskets
Bill	15	40%	6
Amit	40	30%	
Josh	25	60%	
Kevin	8	75%	
Mike	60	25%	
Zheng	44	25%	
André	50	10%	
David	25	20%	
Bob	18	50%	
Lars	15	20%	
Justin	28	25%	

2. On the basis of shooting ability, which five players might you select as the starting lineup for the next basketball game?

_____

3. What other factors might you consider when making this decision?

_____

_____

4. Which player(s) might you encourage to shoot more often? _____

   Why? _____

   _____

5. Which player(s) might you encourage to pass more often? _____

   Why? _____

   _____

Calculating a Discount

Example The list price for a toaster is $45. The toaster is sold at a 12% discount (12% off the list price). What are the savings? (*Reminder:* $12\% = \frac{12}{100} = 0.12$)

Paper and pencil:

$$12\% \text{ of } \$45 \quad = \frac{12}{100} * 45 = \frac{12}{100} * \frac{45}{1}$$

$$= \frac{12 * 45}{100 * 1} = \frac{540}{100}$$

$$= \$5.40$$

Calculator:

Enter 0.12 ⊗ 45 (Enter) ; interpret answer 5.4 as $5.40.

First use your percent sense to estimate the discount for each item in the table below. The **discount** is the amount by which the list price of an item is reduced. It is the amount the customer saves.

Then use your calculator or paper and pencil to calculate the discount. (If necessary, round to the nearest cent.)

Item	List Price	Percent of Discount	Estimated Discount	Calculated Discount
Clock radio	$33.00	20%		
Calculator	$60.00	7%		
Sweater	$20.00	42%		
Tent	$180.00	30%		
Bicycle	$200.00	17%		
Computer	$980.00	25%		
Skis	$325.00	18%		
Double CD	$29.99	15%		
Jacket	$110.00	55%		

Use with Lesson 8.9.

Math Boxes 8.9

1. What is the volume of the cube below?

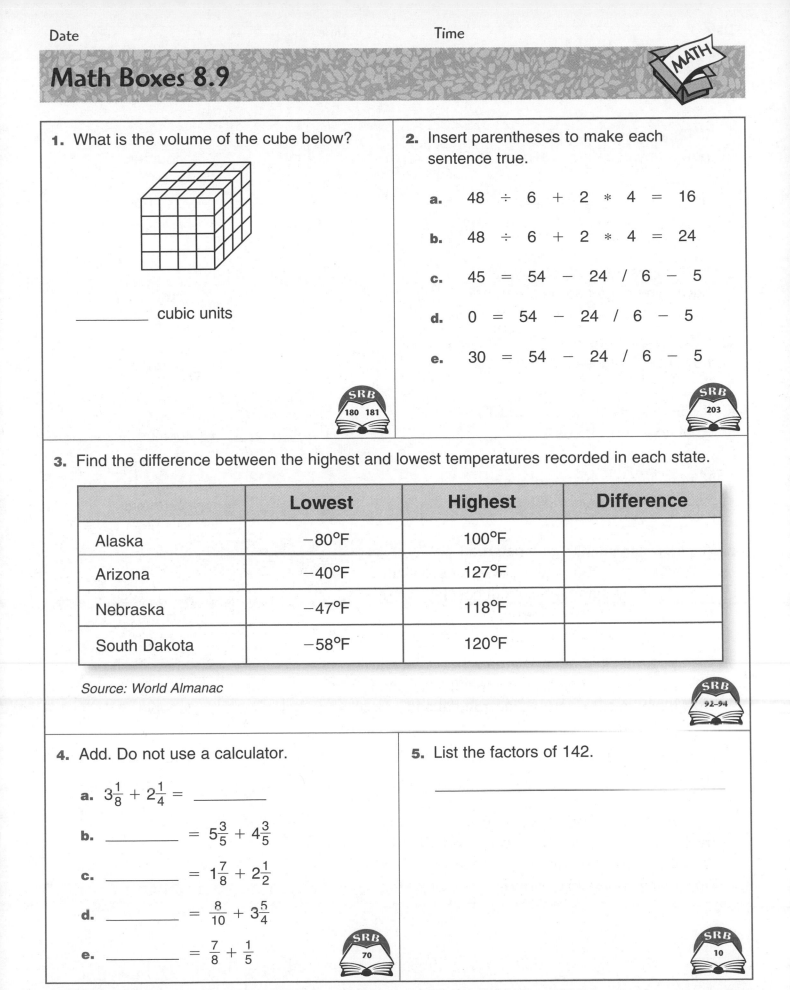

_____ cubic units

2. Insert parentheses to make each sentence true.

a. $48 \div 6 + 2 * 4 = 16$

b. $48 \div 6 + 2 * 4 = 24$

c. $45 = 54 - 24 / 6 - 5$

d. $0 = 54 - 24 / 6 - 5$

e. $30 = 54 - 24 / 6 - 5$

3. Find the difference between the highest and lowest temperatures recorded in each state.

	Lowest	Highest	Difference
Alaska	−80°F	100°F	
Arizona	−40°F	127°F	
Nebraska	−47°F	118°F	
South Dakota	−58°F	120°F	

Source: World Almanac

4. Add. Do not use a calculator.

a. $3\frac{1}{8} + 2\frac{1}{4} =$ _____

b. _____ $= 5\frac{3}{5} + 4\frac{3}{5}$

c. _____ $= 1\frac{7}{8} + 2\frac{1}{2}$

d. _____ $= \frac{8}{10} + 3\frac{5}{4}$

e. _____ $= \frac{7}{8} + \frac{1}{5}$

5. List the factors of 142.

_____

Unit Fractions and Unit Percents

1. If 12 counters are $\frac{1}{5}$ of a set,
   how many counters are in the set?

   _____ counters

2. If 15 counters are $\frac{1}{7}$ of a set,
   how many counters are in the set?

   _____ counters

3. If 31 pages are $\frac{1}{8}$ of a book,
   how many pages are in the book?

   _____ pages

4. If 13 marbles are 1% of the marbles in a jar,
   how many marbles are in the jar?

   _____ marbles

5. If $5.43 is 1% of the cost of a TV,
   what does the TV cost?

   _____ dollars

6. If 84 counters are 10% of a set,
   how many counters are in the set?

   _____ counters

7. After 80 minutes, Dorothy had read
   120 pages of a 300-page book. If she continues
   reading at the same rate, about how long will it
   take her to read the entire book?

   _____ minutes

8. Eighty-four people attended the school concert.
   This was 70% of the number expected to attend.
   How many people were expected to attend?

   _____ people

Challenge

9. In its most recent game, the Lincoln Junior High
   basketball team made 36 baskets, which was
   48% of the shots team members tried.
   How many shots did they try?

   _____ shots

Use with Lesson 8.10.

Using a Unit Fraction or a Unit Percent to Find the Whole

1. Six jars are filled with cookies. The number of cookies in each jar is not known.
   For each clue given below, find the number of cookies in the jar.

Clue	Number of Cookies in Jar
a. $\frac{1}{2}$ jar contains 31 cookies.	
b. $\frac{2}{8}$ jar contains 10 cookies.	
c. $\frac{3}{5}$ jar contains 36 cookies.	
d. $\frac{3}{8}$ jar contains 21 cookies.	
e. $\frac{4}{7}$ jar contains 64 cookies.	
f. $\frac{3}{11}$ jar contains 45 cookies.	

2. Use your percent sense to estimate the list price for each item. Then calculate the
   list price. (*Hint:* First use your calculator to find what 1% is worth.)

Sale Price	Percent of List Price	Estimated List Price	Calculated List Price
$120.00	60%	$180	$200
$100.00	50%		
$8.00	32%		
$255.00	85%		
$77.00	55%		
$80.00	40%		
$9.00	60%		
$112.50	75%		
$450.00	90%		

3. Alan is walking to a friend's house. He covered $\frac{6}{10}$ of the distance in 48 minutes.
   If he continues at the same speed, about how long will the entire walk take?

_____

Using a Unit Fraction or a Unit Percent to Find the Whole (cont.)

4. 24 is $\frac{1}{2}$ of what number? _____

5. $\frac{2}{5}$ is $\frac{1}{2}$ of what number? _____

6. 27 is $\frac{3}{4}$ of what number? _____

7. $\frac{3}{8}$ is $\frac{3}{4}$ of what number? _____

8. 60 is 50% of what number? _____

9. 16 is 25% of what number? _____

10. 40 is 80% of what number? _____

The problems below are from an arithmetic book published in 1906. Solve the problems.

11. If the average coal miner works $\frac{2}{3}$ of a month of 30 days, how many days during the month does he work? _____ days

12. A recipe for fudge calls for $\frac{1}{4}$ of a cake of chocolate. If a cake costs 20¢, find the cost of the chocolate called for by the recipe. _____ ¢

13. In target practice the battleship *Indiana* shot at a target 24 times. If $\frac{3}{4}$ of the shots hit, how many successful shots were fired? _____ shots

14. A collection of mail that required 6 hours for a postman to make with a horse and wagon was made in an automobile in $\frac{5}{12}$ the time. How long did the automobile take? _____ hours

15. How many corks per day does a machine in Spain make from the bark of the cork tree, if it makes $\frac{1}{3}$ of a sack of 15,000 corks in that time? _____ corks

Source: Milne's Progressive Arithmetic

Challenge

16. It's easy to write 1 as a sum of unit fractions if the same unit fraction may be used more than once. *For example:* $\frac{1}{3} + \frac{1}{3} + \frac{1}{6} + \frac{1}{6} = 1$

Try to write 1 as a sum of unit fractions without repeating a fraction. Try to find more than one solution.

_____

Use with Lesson 8.10.

Math Boxes 8.10

1. Use your Geometry Template to draw a parallelogram.

What are some other names for the figure you drew?

_____

_____

SRB
135 136

2. Write > or <.

a. 50% _____ $\frac{2}{3}$

b. 620 − 80 _____ 30 * 40

c. $\frac{7}{8}$ _____ $\frac{1}{4} + \frac{2}{4}$

d. 20 * 19 _____ 20^2

e. 0.35 + 0.25 _____ $\frac{1}{8} + \frac{1}{8}$

3. Draw and label a 30° angle.

SRB
190

4. Circle the numbers below that are evenly divisible by 6.

148   293   762   1,050   984

SRB
11

5. Solve.

a. 75
   * 88

b. 425
   * 68

c. 759
   * 13

d. 422
   * 185

SRB
19 20

Math Boxes 8.11

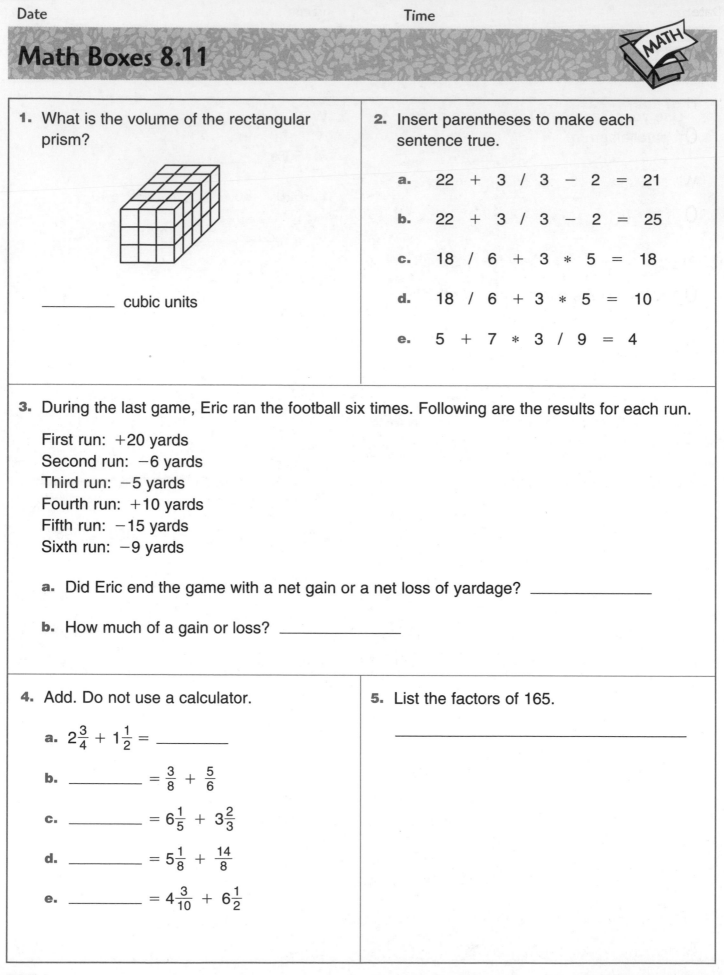

1. What is the volume of the rectangular prism?

_____ cubic units

2. Insert parentheses to make each sentence true.

a. 22 + 3 / 3 − 2 = 21

b. 22 + 3 / 3 − 2 = 25

c. 18 / 6 + 3 * 5 = 18

d. 18 / 6 + 3 * 5 = 10

e. 5 + 7 * 3 / 9 = 4

3. During the last game, Eric ran the football six times. Following are the results for each run.

First run: +20 yards
Second run: −6 yards
Third run: −5 yards
Fourth run: +10 yards
Fifth run: −15 yards
Sixth run: −9 yards

a. Did Eric end the game with a net gain or a net loss of yardage? _____

b. How much of a gain or loss? _____

4. Add. Do not use a calculator.

a. $2\frac{3}{4} + 1\frac{1}{2} =$ _____

b. _____ $= \frac{3}{8} + \frac{5}{6}$

c. _____ $= 6\frac{1}{5} + 3\frac{2}{3}$

d. _____ $= 5\frac{1}{8} + \frac{14}{8}$

e. _____ $= 4\frac{3}{10} + 6\frac{1}{2}$

5. List the factors of 165.

_____

Class Survey

1. How many people live in your home?

   O 1–2 people   O 3–5 people   O 6 or more people

2. What language do you speak at home?

   O English   O Spanish   O Other: _____

3. Are you right- or left-handed?

   O right-handed   O left-handed

4. How long have you lived at your current address? (Round to the nearest year.)

   _____ years

5. Pick one of the questions above. Tell why someone you don't know might be
   interested in your answer to the question you picked.

   _____

   _____

   _____

   _____

   _____

6. Fifteen percent of the 20 students in Ms. Swanson's class were left-handed.

   How many students were left-handed?   _____ students

7. About 85% of the 600 students at Emerson Middle School speak English at home.
   Another 10% speak Spanish, and 5% speak other languages. About how many
   students speak each language at home?

   English: _____ students

   Spanish: _____ students

   Other: _____ students

8. The government reported that 5% of 90,000,000 workers do not have jobs.

   How many workers were jobless? _____ workers

Rural and Urban Populations

The U.S. Census Bureau classifies where people live according to the following rule: **Rural** areas are communities having fewer than 2,500 people each. **Urban** areas are communities having 2,500 or more people each.

1. According to the Census Bureau's definition, do you live in a rural or an urban area?

   _____

   How did you decide? _____

   _____

   _____

   _____

Today more than three out of every four residents in the United States live in areas the Census Bureau defines as urban. This was not always the case. When the United States was formed, it was a rural nation.

Work with your classmates and use the information in the *Student Reference Book,* pages 308, 309, and 334 to examine the transformation of the United States from a rural to an urban nation.

2. My group is to estimate the number of people living in _____ areas in
   _____ (rural or urban)
   _____ .
   (1790, 1850, 1900, or 2000)

3. The total U.S. population in _____ was _____ .
   (1790, 1850, 1900, or 2000)

4. Estimate: The number of people living in _____ areas in
   _____ (rural or urban)
   _____ was about _____ .
   (1790, 1850, 1900, or 2000)
   Make sure your answer is rounded to the nearest 100,000.

5. Our estimation strategy was _____

   _____

   _____

   _____ .

Rural and Urban Populations (cont.)

6. Use the estimates from the groups in your class to complete the following table.

Estimated Rural and Urban Populations, 1790–2000		
Year	Estimated Rural Population	Estimated Urban Population
1790		
1850		
1900		
2000		

7. Is it fair to say that for more than half our nation's history, the **majority** of the population lived in rural areas?

Explain your answer.

Vocabulary
majority means *more than one-half of a count*

_____

_____

_____

_____

8. About how many times larger was the rural population in 2000 than in 1790?

_____

9. About how many times larger was the urban population in 2000 than in 1790?

_____

Challenge

10. In which decade do you think the urban population became larger than the rural population?

_____

_____

Division

Math Message

1. How many 2-pound boxes of candy can be made from 10 pounds of candy?

   _____ boxes

2. How many $\frac{3}{4}$-pound boxes of candy can be made from 6 pounds of candy?

   _____ boxes

3. Sam has 5 pounds of peanut brittle. He wants to pack it in $\frac{3}{4}$-pound packages.

   How many full packages can he make? _____ full packages

   Will any peanut brittle be left over? _____   How much? _____ pound

4.
```
 ┌──────────────────────────────────────────────────────────
 │0      1      2      3      4      5      6
 │ INCHES
```

   a. How many 2-inch segments are there in 6 inches? _____ segments

   b. How many $\frac{1}{2}$-inch segments are there in 6 inches? _____ segments

   c. How many $\frac{1}{8}$-inch segments are there in $\frac{3}{4}$ of an inch? _____ segments

---

Common Denominator Division

One way to divide fractions uses common denominators:

Step 1 Rename the fractions using a common denominator.
Step 2 Divide the numerators.

This method can also be used for whole or mixed numbers divided by fractions.

Examples

$3 \div \frac{3}{4} = ?$	$\frac{1}{3} \div \frac{1}{6} = ?$	$3\frac{3}{5} \div \frac{3}{5} = \frac{18}{5} \div \frac{3}{5}$
$3 \div \frac{3}{4} = \frac{12}{4} \div \frac{3}{4}$	$\frac{1}{3} \div \frac{1}{6} = \frac{2}{6} \div \frac{1}{6}$	$= 18 \div 3 = 6$
$= 12 \div 3 = 4$	$= 2 \div 1 = 2$	

Common Denominator Division (cont.)

Solve.

1. $4 \div \frac{4}{5} =$ _____

2. $\frac{5}{6} \div \frac{1}{18} =$ _____

3. $3\frac{1}{3} \div \frac{5}{6} =$ _____

4. $6\frac{3}{5} \div 2\frac{2}{10} =$ _____

5. $2 \div \frac{2}{5} =$ _____

6. $2 \div \frac{2}{3} =$ _____

7. $6 \div \frac{3}{5} =$ _____

8. $\frac{1}{2} \div \frac{1}{8} =$ _____

9. $\frac{3}{5} \div \frac{1}{10} =$ _____

10. $\frac{6}{5} \div \frac{3}{10} =$ _____

11. $1\frac{1}{2} \div \frac{3}{4} =$ _____

12. $4\frac{1}{5} \div \frac{3}{5} =$ _____

13. Chase is packing cookies in $\frac{1}{2}$-pound bags. He has 10 pounds of cookies.

How many bags can he pack? _____ bags

14. Regina is cutting lanyard to make bracelets. She has 15 feet of lanyard and

needs $1\frac{1}{2}$ feet for each bracelet. How many bracelets can she make? _____ bracelets

15. Eric is planning a pizza party. He has 3 large pizzas. He figures each person will eat $\frac{3}{8}$ of a pizza. How many people can attend the party, including himself?

_____ people

Math Boxes 8.12

1. Use your Geometry Template to draw a triangle.

What kind of triangle did you draw?

_____

2. Write > or <.

a. $15 + 28$ _____ 10^2

b. $40 + 40$ _____ $3 * 30$

c. $\frac{1}{2} + \frac{1}{2}$ _____ $\frac{3}{4}$

d. $\frac{19}{20}$ _____ $0.6 + 0.3$

e. $55 \div 5$ _____ $120 \div 12$

3. Draw and label a 170° angle.

4. Circle the numbers below that are evenly divisible by 9.

3,735   2,043   192   769   594

5. Solve.

a.   429
     $* 15$

b.   134
     $* 82$

c.   706
     $* 189$

Use with Lesson 8.12.

Time to Reflect

1. Describe several situations in which knowing how to solve percent-of problems would be helpful.

_____

_____

_____

_____

_____

2. Explain why you think it is important to know how to find equivalent fractions with common denominators.

_____

_____

_____

_____

_____

3. Look back through your journal pages for this unit. What do you think is the most important skill or concept you learned in this unit? Explain why.

_____

_____

_____

_____

Math Boxes 8.13

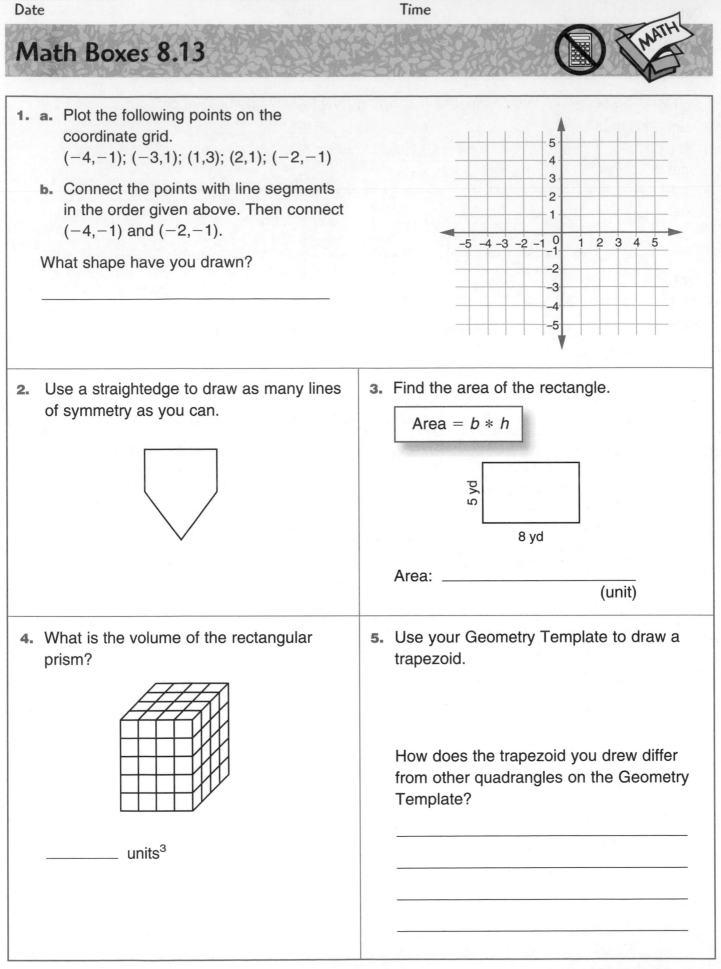

1. **a.** Plot the following points on the coordinate grid.

    $(-4,-1)$; $(-3,1)$; $(1,3)$; $(2,1)$; $(-2,-1)$

    b. Connect the points with line segments in the order given above. Then connect $(-4,-1)$ and $(-2,-1)$.

    What shape have you drawn?

    _____

2. Use a straightedge to draw as many lines of symmetry as you can.

3. Find the area of the rectangle.

    Area = $b * h$

    5 yd

    8 yd

    Area: _____

    (unit)

4. What is the volume of the rectangular prism?

    _____ units³

5. Use your Geometry Template to draw a trapezoid.

    How does the trapezoid you drew differ from other quadrangles on the Geometry Template?

    _____

    _____

    _____

    _____

Use with Lesson 8.13.

Plotting a Turtle

Points on a coordinate grid are named by ordered number pairs. The first number in an ordered number pair locates the point along the horizontal axis. The second number locates the point along the vertical axis. To mark a point on a coordinate grid, first go right (or left) on the horizontal axis. Then go up (or down) from there.

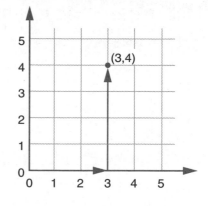

Plot an outline of the turtle on the graph below. Start with the nose, at point (8,12).

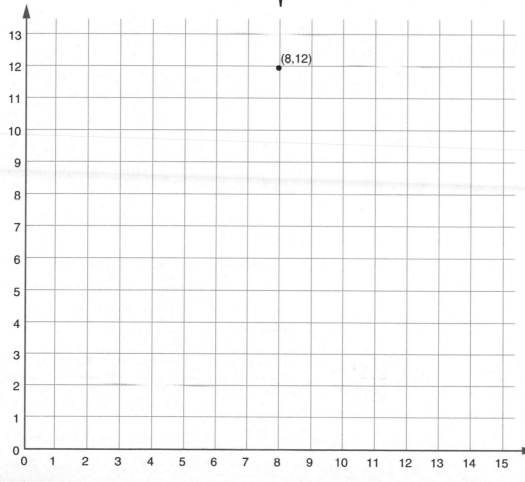

Hidden Treasure Gameboards

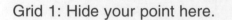

Each player uses Grids 1 and 2.

Grid 1: Hide your point here.

Grid 2: Guess other player's point here.

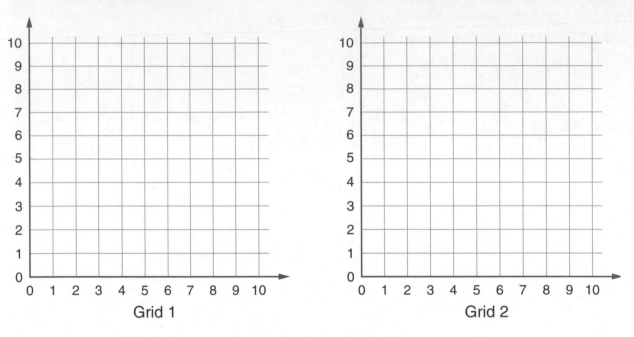

Grid 1

Grid 2

Use Grids 1 and 2 to play another game.

Grid 1: Hide your point here.

Grid 2: Guess other player's point here.

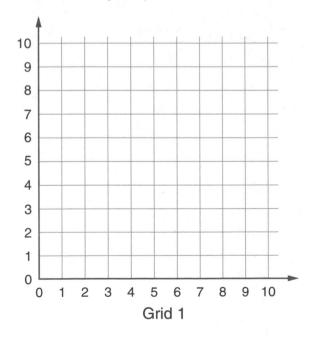

Grid 1

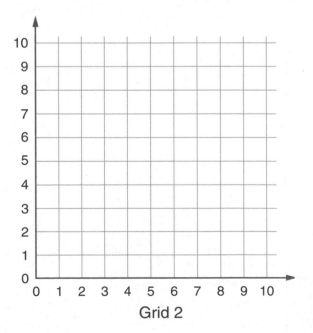

Grid 2

Use with Lesson 9.1.

Matching Graphs to Number Stories

1. Draw a line matching each graph below to the number story that it best fits.

   a. Juanita started with $350. She saved
      another $25 every week.

Graph A

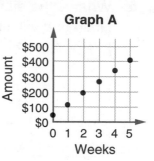

   b. Meredith received $350 for her birthday.
      She deposited the entire amount in the
      bank. Every week she withdrew $50.

Graph B

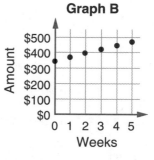

   c. Julian started a new savings account
      with $50. Every week after that he
      deposited $75.

Graph C

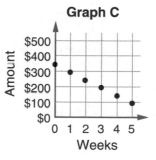

2. Explain how you decided which graph matches each number story.

   _____

   _____

   _____

   _____

3. Circle the rule below that best fits the number story in Problem 1a above.

   Savings = $350 + (25 * number of weeks)

   Savings = $350 − (25 * number of weeks)

   Savings = $350 * number of weeks

Math Boxes 9.1

1. a. What is the diameter of the largest circle that will fit inside the box for this problem?

_____

b. Explain your answer.

_____

_____

_____

SRB 143

2. a. Write the largest number you can that is less than 1 by using each of the following digits only once: 4 7 5 2.

_____

b. Write the number in words.

_____

_____

_____

SRB 30

3. Write a number model that describes each of the shaded rectangles.

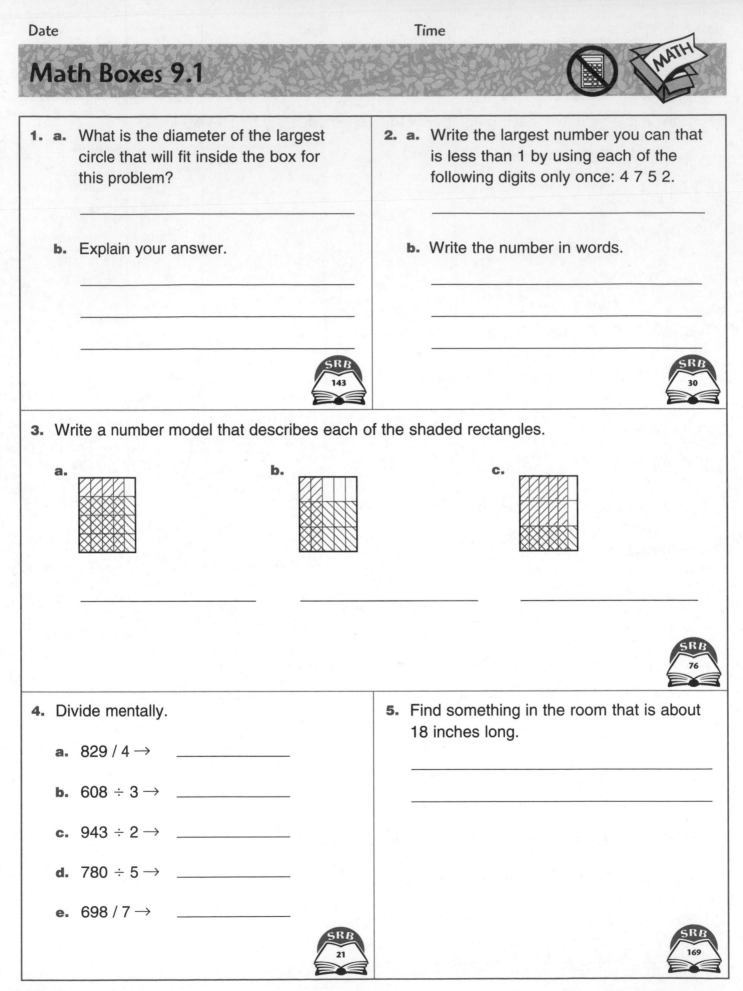

a.

_____

b.

_____

c.

_____

SRB 76

4. Divide mentally.

a. 829 / 4 → _____

b. 608 ÷ 3 → _____

c. 943 ÷ 2 → _____

d. 780 ÷ 5 → _____

e. 698 / 7 → _____

SRB 21

5. Find something in the room that is about 18 inches long.

_____

_____

SRB 169

Use with Lesson 9.1.

Plotting a Map

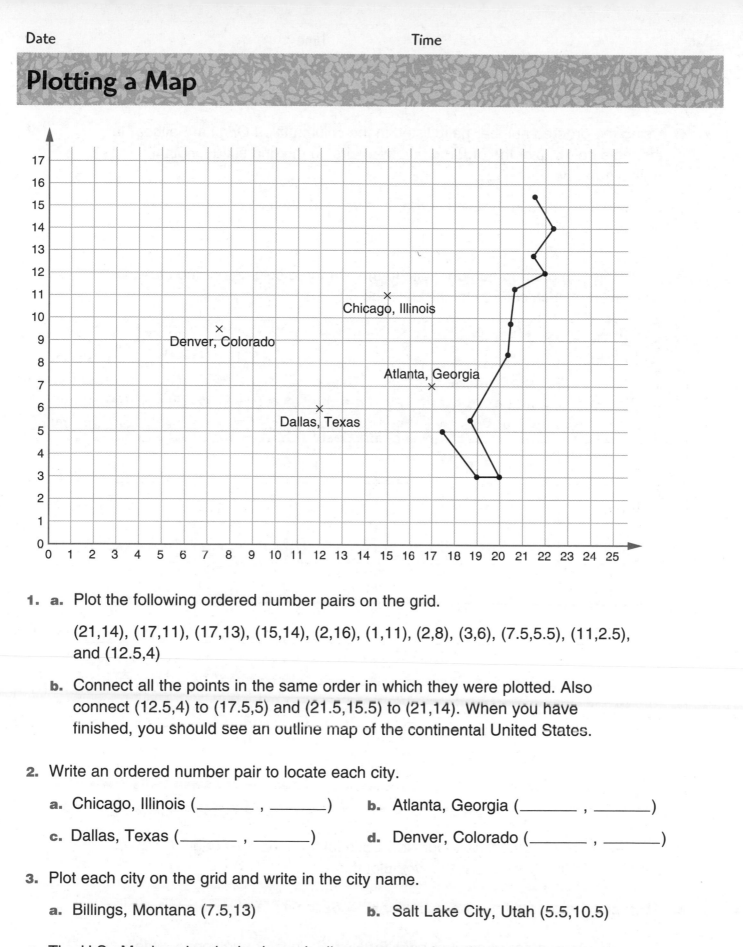

1. **a.** Plot the following ordered number pairs on the grid.

   (21,14), (17,11), (17,13), (15,14), (2,16), (1,11), (2,8), (3,6), (7.5,5.5), (11,2.5), and (12.5,4)

   b. Connect all the points in the same order in which they were plotted. Also connect (12.5,4) to (17.5,5) and (21.5,15.5) to (21,14). When you have finished, you should see an outline map of the continental United States.

2. Write an ordered number pair to locate each city.

   a. Chicago, Illinois (_____ , _____)   **b.** Atlanta, Georgia (_____ , _____)

   c. Dallas, Texas (_____ , _____)   **d.** Denver, Colorado (_____ , _____)

3. Plot each city on the grid and write in the city name.

   a. Billings, Montana (7.5,13)   **b.** Salt Lake City, Utah (5.5,10.5)

4. The U.S.–Mexican border is shown by line segments from (3,6) to (7.5,5.5) and from (7.5,5.5) to (11,2.5). Write the border name on the grid.

Sailboat Graphs

1. **a.** Using the ordered number pairs listed in the column titled Original Sailboat in the table below, plot the ordered number pairs on the grid titled Original Sailboat on the next page.

   b. Connect the points in the same order that you plot them. You should see the outline of a sailboat.

2. Fill in the missing ordered number pairs in the last three columns of the table. Use the rule given in each column to calculate the ordered number pairs.

Original Sailboat	New Sailboat 1 Rule: Double each number of the original pair.	New Sailboat 2 Rule: Double the first number of the original pair.	New Sailboat 3 Rule: Double the second number of the original pair.
(8,1)	(16, 2)	(16, 1)	(8, 2)
(5,1)	(10, 2)	(10, 1)	(5, 2)
(5,7)	(10, 14)	(10, 7)	(5, 14)
(1,2)	(_____ , _____)	(_____ , _____)	(_____ , _____)
(5,1)	(_____ , _____)	(_____ , _____)	(_____ , _____)
(0,1)	(_____ , _____)	(_____ , _____)	(_____ , _____)
(2,0)	(_____ , _____)	(_____ , _____)	(_____ , _____)
(7,0)	(_____ , _____)	(_____ , _____)	(_____ , _____)
(8,1)	(_____ , _____)	(_____ , _____)	(_____ , _____)

3. **a.** Plot the ordered number pairs for New Sailboat 1 on the next page. Connect the points in the same order that you plot them.

   b. Then plot the ordered number pairs for New Sailboat 2 and connect the points.

   c. Finally, plot the ordered number pairs for New Sailboat 3 and connect the points.

Use with Lesson 9.2.

Sailboat Graphs (cont.)

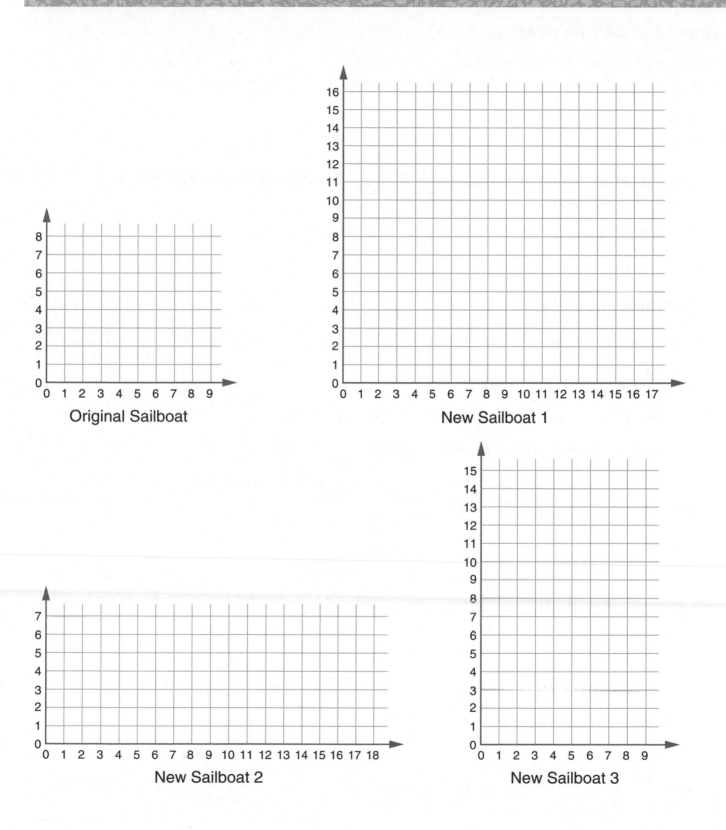

Original Sailboat

New Sailboat 1

New Sailboat 2

New Sailboat 3

Place-Value Puzzles

1. The digit in the thousands place is 8.

   The digit in the ones place is the sum of the digits in three centuries. (*Hint:* If there are _____ years in one century, then there are _____ years in three centuries.)

   The digit in the millions place is $\frac{1}{10}$ of 40.

   The digit in the hundred-thousands place is $\frac{1}{2}$ of the digit in the thousands place.

   The digit in the hundreds place is the sum of the digit in the millions place and the digit in the ones place.

   The rest of the digits are all 5s.

   _____ _____ , _____ _____ _____ , _____ _____ _____

   Write this number in words. _____

   _____

2. The digit in the tenths place is 1.

   The digit in the ones place is double the digit in the tenths place.

   The digit in the thousandths place is $\frac{1}{3}$ of 21.

   The digit in the hundreds place is three times the digit in the tenths place.

   The digit in the ten-thousands place is an odd number less than 6 that you haven't used yet.

   The rest of the digits are all 9s.

   _____ _____ , _____ _____ _____ . _____ _____ _____

   Write this number in words. _____

   _____

3. Make up a puzzle of your own.

   _____

   _____

   _____

   _____

Use with Lesson 9.2.

Math Boxes 9.2

1. Complete the "What's My Rule?" table and state the rule.

Rule: _____

in	out
2	−10
	0
16	4
3	
	−5

SRB 215 216

2. Complete the table.

Standard Notation	Scientific Notation
300	$3 * 10^2$
3,000	$3 * 10^3$
4,000	
500	
	$7 * 10^3$

SRB 8

3. Multiply. Show your work.

a.   55
     * 37

b.   92
     * 74

c.   318
     * 64

SRB 19 20

4. Measure each angle.

a. ∠*LID* measures

about _____ °.

b. ∠*FUN* measures

about _____ °.

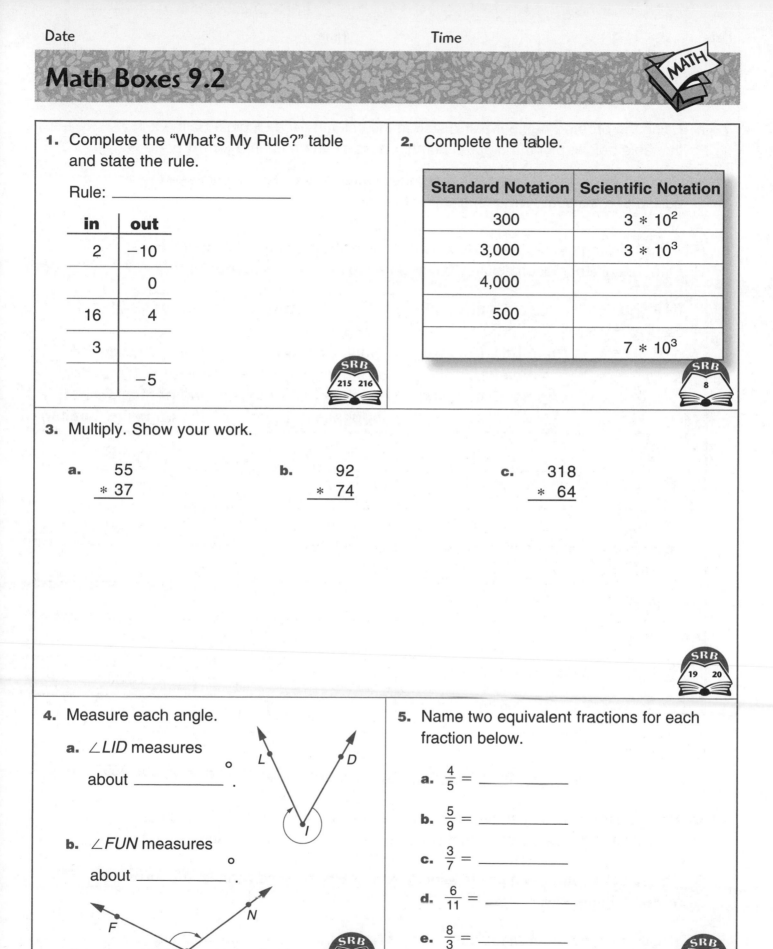

SRB 188–189

5. Name two equivalent fractions for each fraction below.

a. $\frac{4}{5}$ = _____

b. $\frac{5}{9}$ = _____

c. $\frac{3}{7}$ = _____

d. $\frac{6}{11}$ = _____

e. $\frac{8}{3}$ = _____

SRB 59 60

More Sailboat Graphs

1. **a.** Using the ordered number pairs listed in the column titled Original Sailboat in the table below, plot the ordered number pairs on the grid on the next page.

   b. Connect the points in the same order they were plotted. You should see the outline of a sailboat. Write "original" in the sail.

2. Fill in the missing ordered number pairs in the last three columns of the table. Use the rule given in each column to calculate the ordered number pairs.

Original Sailboat	New Sailboat 1 Rule: Add 10 to the first number of the original number pair.	New Sailboat 2 Rule: Change the first number of the original pair to the opposite number.	New Sailboat 3 Rule: Change the second number of the original pair to the opposite number.
(9,3)	(19, 3)	(−9, 3)	(9, −3)
(6,3)	(16, 3)	(−6, 3)	(6, −3)
(6,9)	(16, 9)	(−6, 9)	(6, −9)
(2,4)	(____ , ____)	(____ , ____)	(____ , ____)
(6,3)	(____ , ____)	(____ , ____)	(____ , ____)
(1,3)	(____ , ____)	(____ , ____)	(____ , ____)
(3,2)	(____ , ____)	(____ , ____)	(____ , ____)
(8,2)	(____ , ____)	(____ , ____)	(____ , ____)
(9,3)	(____ , ____)	(____ , ____)	(____ , ____)

3. **a.** Plot the ordered number pairs for New Sailboat 1 on the next page. Connect the points in the same order that you plot them. Write "1" in the sail.

   b. Then plot the ordered number pairs for New Sailboat 2 and connect the points. Write "2" in the sail.

   c. Finally, plot the ordered number pairs for New Sailboat 3 and connect the points. Write "3" in the sail.

Use with Lesson 9.3.

More Sailboat Graphs (cont.)

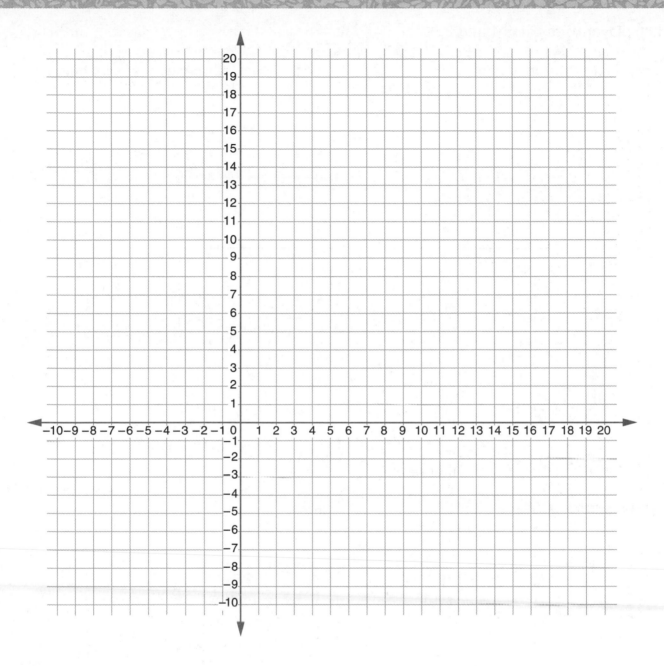

4. Use the following rule to create a new sailboat figure on the grid above.
   Label it "4."

   Rule: Add 10 to the second number of the original pair. Leave the first
   number unchanged.

   Try to plot the new coordinates without listing them.

Advanced Hidden Treasure Gameboards

Each player uses Grids 1 and 2.

Grid 1: Hide your point here.

Grid 2: Guess other player's point here.

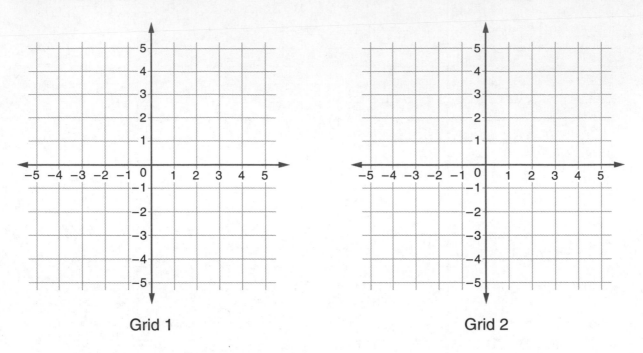

Grid 1

Grid 2

Use Grids 1 and 2 to play another game.

Grid 1: Hide your point here.

Grid 2: Guess other player's point here.

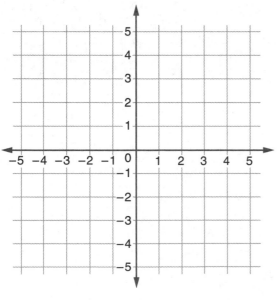

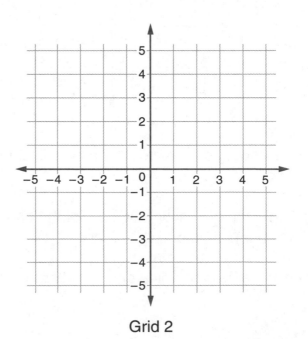

Grid 1

Grid 2

Use with Lesson 9.3.

Math Boxes 9.3

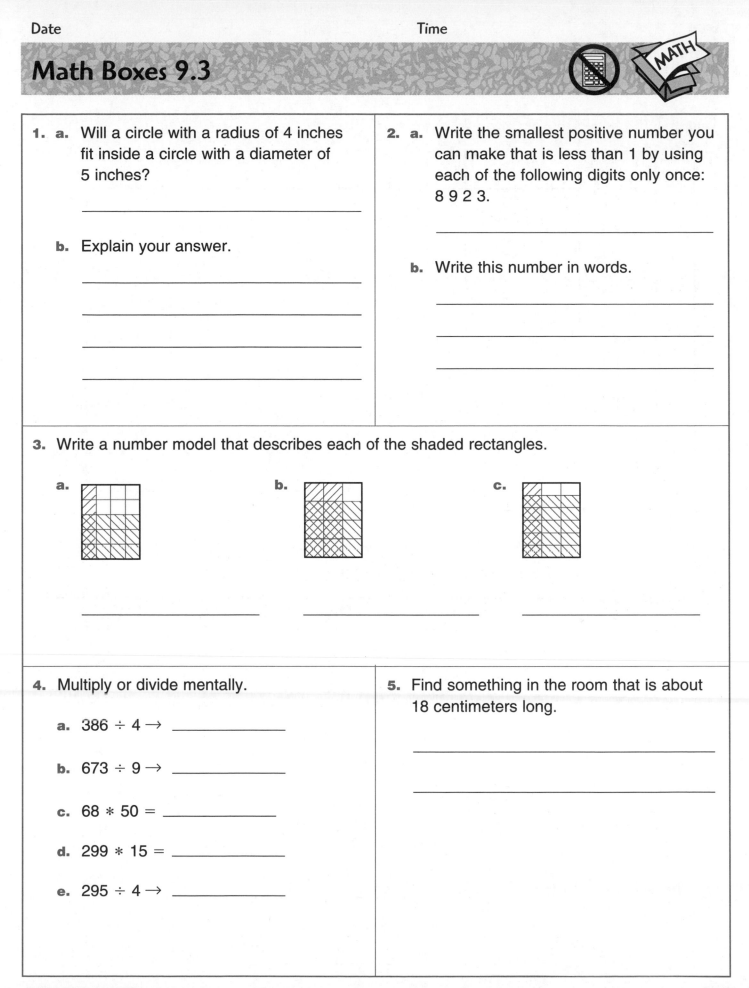

1. a. Will a circle with a radius of 4 inches fit inside a circle with a diameter of 5 inches?

_____

b. Explain your answer.

_____

_____

_____

_____

2. a. Write the smallest positive number you can make that is less than 1 by using each of the following digits only once: 8 9 2 3.

_____

b. Write this number in words.

_____

_____

_____

3. Write a number model that describes each of the shaded rectangles.

a.

b.

c.

_____   _____   _____

4. Multiply or divide mentally.

a. $386 \div 4 \rightarrow$ _____

b. $673 \div 9 \rightarrow$ _____

c. $68 * 50 =$ _____

d. $299 * 15 =$ _____

e. $295 \div 4 \rightarrow$ _____

5. Find something in the room that is about 18 centimeters long.

_____

_____

Areas of Rectangles

1 cm²   1 cm

1 cm

base (or length)

height (or width)

B

A

C

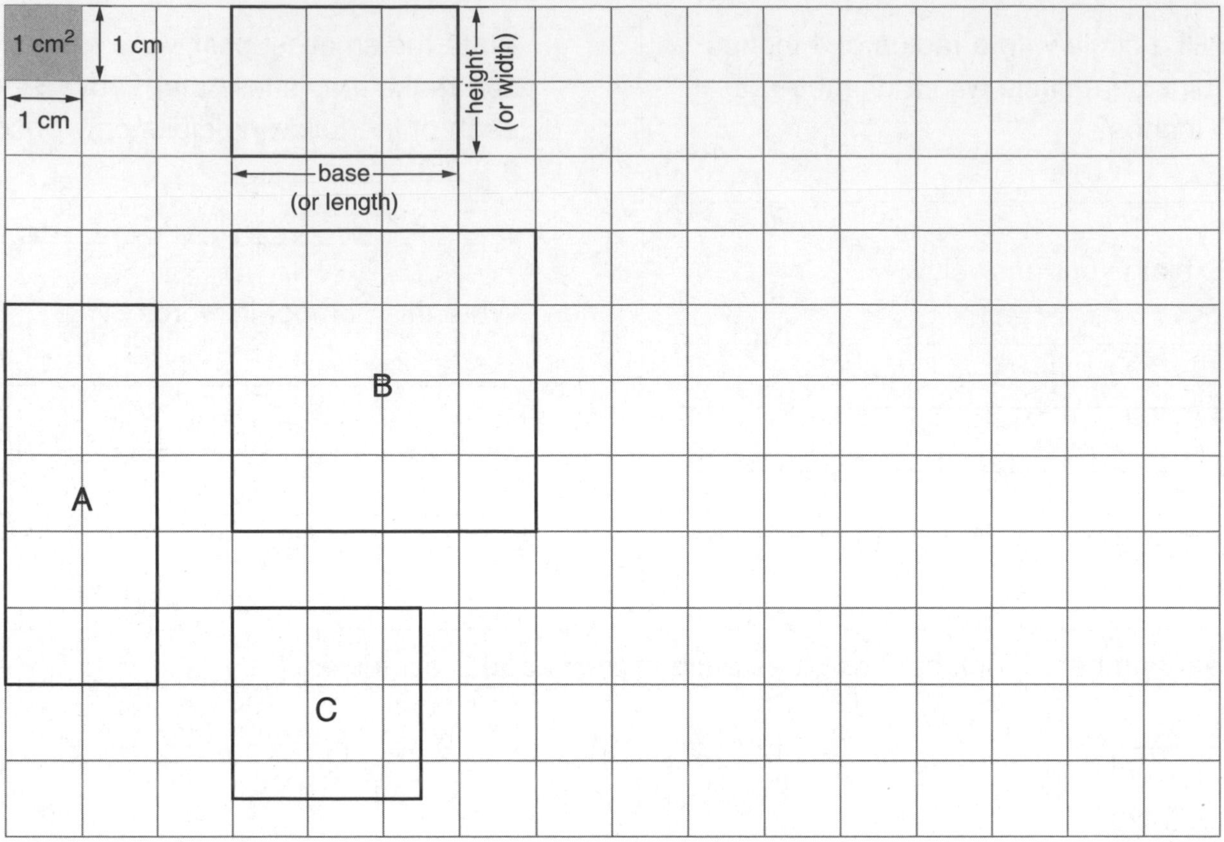

1. Fill in the table. Draw rectangles D, E, and F on the grid.

Rectangle	Base (length)	Height (width)	Area
A	_____ cm	_____ cm	_____ cm²
B	_____ cm	_____ cm	_____ cm²
C	_____ cm	_____ cm	_____ cm²
D	6 cm	_____ cm	12 cm²
E	3.5 cm	_____ cm	14 cm²
F	3 cm	_____ cm	10.5 cm²

2. Write a formula for finding the area of a rectangle.

Area = _____

     Use with Lesson 9.4.

Date _____ Time _____

Area Problems

1. A bedroom floor is 12 feet by 15 feet (4 yards by 5 yards).

   Floor area = _____ square feet

   Floor area = _____ square yards

   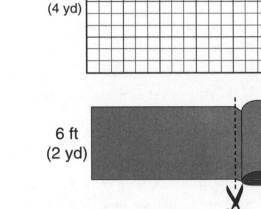
   15 ft (5 yd)
   12 ft (4 yd)

2. Imagine that you want to buy carpet for the bedroom in Problem 1. The carpet comes on a roll that is 6 feet (2 yards) wide. The carpet salesperson unrolls the carpet to the length you want and cuts off your piece. How long a piece will you need to cover the bedroom floor? _____

   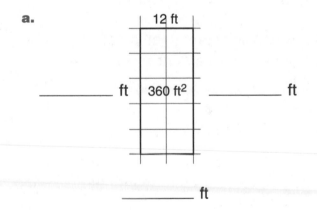
   6 ft (2 yd)

---

3. Calculate the areas for the figures below.

   a.
   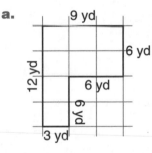
   9 yd
   6 yd
   12 yd
   6 yd
   6 yd
   3 yd

   Area = _____ yd²

   b.
   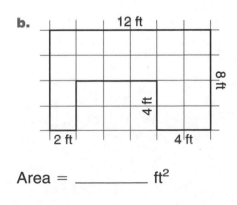
   12 ft
   8 ft
   4 ft
   2 ft
   4 ft

   Area = _____ ft²

4. Fill in the missing lengths for the figures below.

   a.
   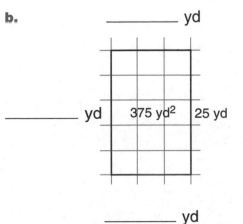
   12 ft
   _____ ft   360 ft²   _____ ft

   _____ ft

   b.
   _____ yd

   _____ yd   375 yd²   25 yd

   _____ yd

Review of 2-Dimensional Figures

Match each description of a geometric figure in Column A with its name in Column B.
Not every name in Column B has a match.

A

B

a. A polygon with 4 right angles
   and 4 sides of the same length

_____ octagon

_____ rhombus

b. A polygon with 4 sides, no two
   of which need to be the same
   size

_____ right angle

_____ acute angle

c. A quadrilateral with exactly one
   pair of opposite sides that are
   parallel

_____ trapezoid

d. Lines in the same plane that
   never intersect

_____ hexagon

_____ square

e. A parallelogram (that is not a
   square) with all sides the same
   length

_____ equilateral triangle

_____ perpendicular lines

f. A polygon with 8 sides

_____ parallel lines

g. Two intersecting lines that form
   a right angle

_____ pentagon

h. A polygon with 5 sides

_____ isosceles triangle

i. An angle that measures 90°

_____ quadrilateral

j. A triangle with all sides the
   same length

Use with Lesson 9.4.

Math Boxes 9.4

MATH

1. Complete the "What's My Rule?" table and state the rule.

Rule: _____

in	out
5	
3	−2
	5
0	
	−7

2. Complete the table.

Standard Notation	Scientific Notation
8,000	$8 * 10^3$
60,000	
	$5 * 10^5$
	$4 * 10^5$
700,000	

3. Divide. Show your work.

a. _____ → 384 ÷ 21   **b.** 2,935 ÷ 17 → _____   **c.** 8,796 ÷ 43 → _____

4. What is the measure of ∠A?

_____

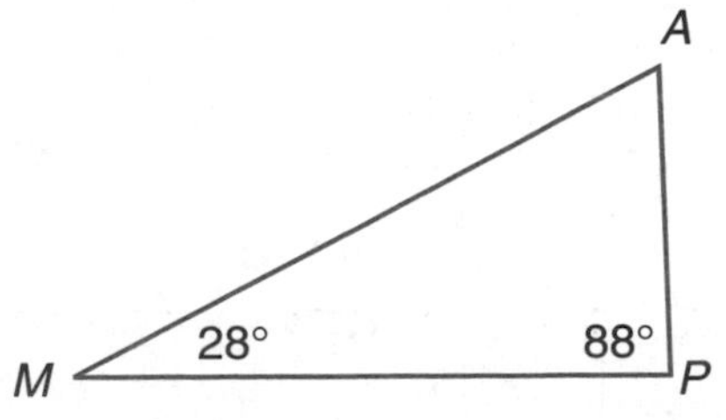

5. Name two equivalent fractions for each fraction below.

a. $\frac{7}{8}$ = _____

b. $\frac{3}{10}$ = _____

c. $\frac{6}{7}$ = _____

d. $\frac{1}{6}$ = _____

e. $\frac{12}{5}$ = _____

Personal References

In *Fourth Grade Everyday Mathematics,* you found **personal references** for metric and U.S. customary units of length, weight, and capacity. These references are familiar objects whose sizes approximate standard measures. For example, for many people the distance across the tip of their smallest finger is about 1 centimeter.

Now you are working with **area,** so try to find personal references for area units.

Spend some time searching through your workspace or classroom to find common objects that have areas of 1 square inch, 1 square foot, 1 square yard, 1 square centimeter, and 1 square meter. The areas do not have to be exact, but they should be reasonable estimates. Ask a friend to look for references with you. Try to find more than one reference for each measure.

Personal References for Common Units of Area

Unit	My Personal References
1 square inch (1 in.^{2})	
1 square foot (1 ft^2)	
1 square yard (1 yd^2)	
1 square centimeter (1 cm^2)	
1 square meter (1 m^2)	

Use with Lesson 9.5.

Finding Areas of Nonrectangular Figures

In the previous lesson, you calculated the areas of rectangular figures using two different methods.

- You counted the total number of unit squares and parts of unit squares that fit neatly inside the figure.

- You used the formula $A = b * h$, where the letter A stands for *area*, the letter b for the length of the *base*, and the letter h for the *height*.

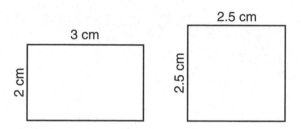

However, many times you will need to find the area of a figure that is not a rectangle. Unit squares will not fit neatly inside the figure, and you won't be able to use the formula for the area of a rectangle.

Working with a partner, think of a way to find the area of each of the figures below.

1. What is the area of triangle *ABC*?

_____

2. What is the area of triangle *XYZ*?

_____

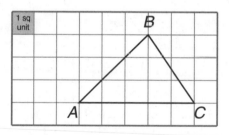

3. What is the area of parallelogram *GRAM*?

_____

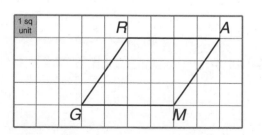

Areas of Triangles and Parallelograms

Use the rectangle method to find the area of each triangle and parallelogram below.

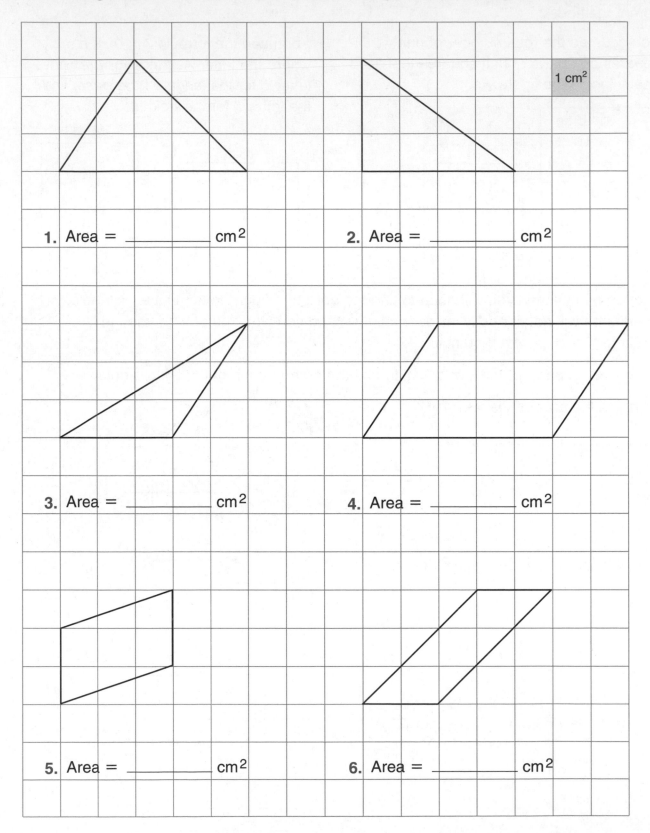

1 cm^2

1. Area = _____ cm^2

2. Area = _____ cm^2

3. Area = _____ cm^2

4. Area = _____ cm^2

5. Area = _____ cm^2

6. Area = _____ cm^2

Math Boxes 9.5

1. Plot and label the ordered number pairs on the grid.

M: (2,5)

N: (−2,1)

O: (−3,−4)

P: (−4,3)

Q: (6,−2)

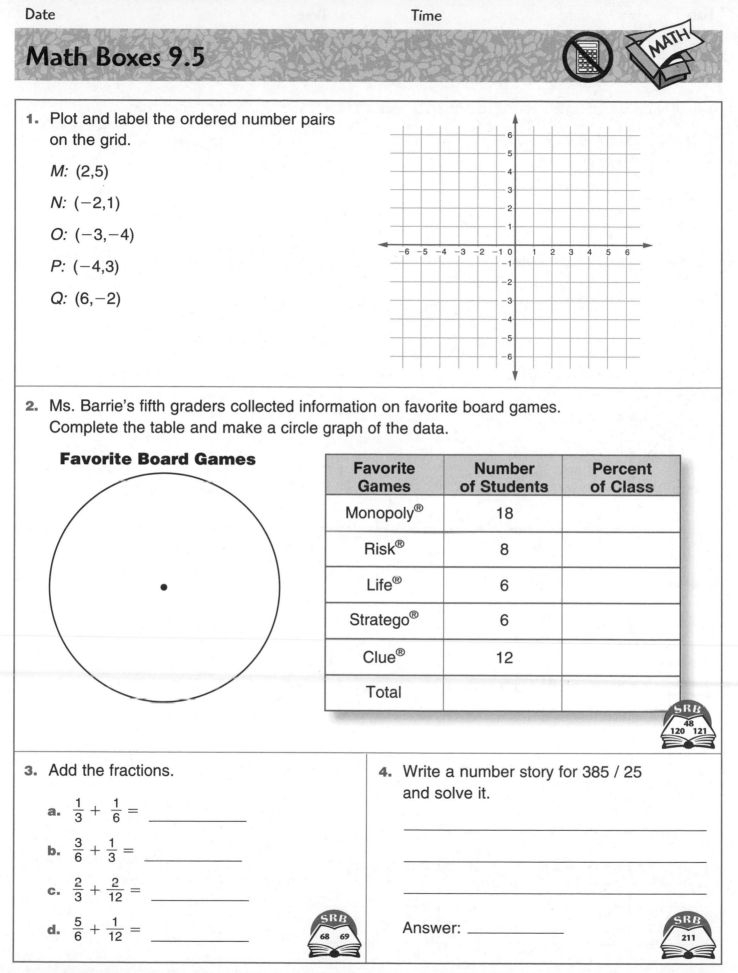

2. Ms. Barrie's fifth graders collected information on favorite board games. Complete the table and make a circle graph of the data.

Favorite Board Games

Favorite Games	Number of Students	Percent of Class
Monopoly®	18	
Risk®	8	
Life®	6	
Stratego®	6	
Clue®	12	
Total		

SRB
48
120 121

3. Add the fractions.

a. $\frac{1}{3} + \frac{1}{6} =$ _____

b. $\frac{3}{6} + \frac{1}{3} =$ _____

c. $\frac{2}{3} + \frac{2}{12} =$ _____

d. $\frac{5}{6} + \frac{1}{12} =$ _____

SRB
68 69

4. Write a number story for 385 / 25 and solve it.

_____

_____

_____

Answer: _____

SRB
211

The Rectangle Method

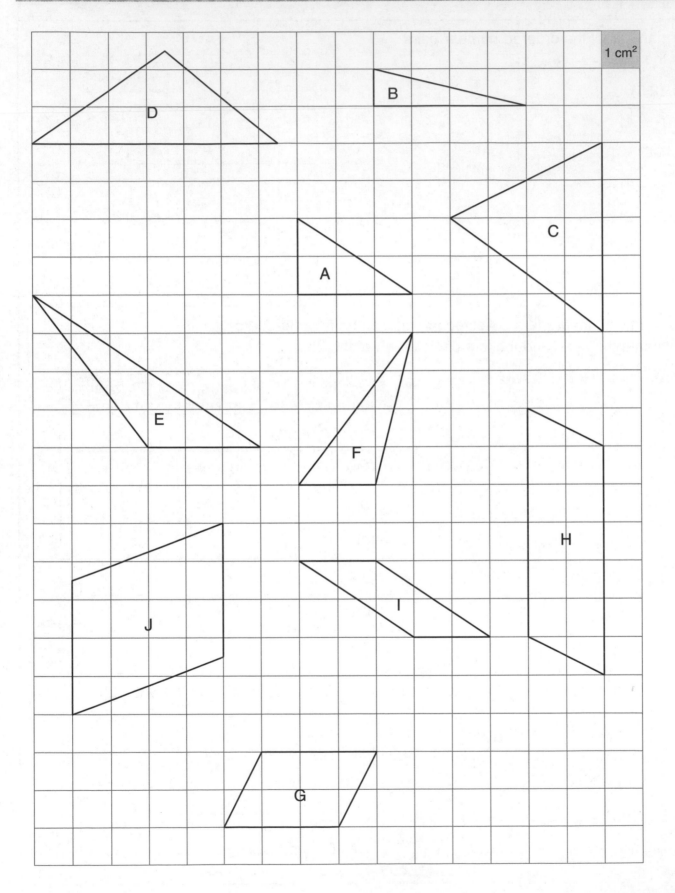

1 cm²

D

B

C

A

E

F

H

J

I

G

Use with Lesson 9.6.

Finding Areas of Triangles and Parallelograms

1. Fill in the table. All figures are shown on journal page 318.

	Area	base	height	base * height
Triangles				
A	3 cm²	3 cm	2 cm	6 cm²
B	_____ cm²	_____ cm	_____ cm	_____ cm²
C	_____ cm²	_____ cm	_____ cm	_____ cm²
D	_____ cm²	_____ cm	_____ cm	_____ cm²
E	_____ cm²	3 cm	4 cm	_____ cm²
F	_____ cm²	_____ cm	_____ cm	_____ cm²
Parallelograms				
G	6 cm²	3 cm	2 cm	6 cm²
H	_____ cm²	_____ cm	_____ cm	_____ cm²
I	_____ cm²	_____ cm	2 cm	_____ cm²
J	_____ cm²	_____ cm	_____ cm	_____ cm²

2. Examine the results of Figures A–F. Propose a formula for the area of a triangle as an equation and as a word sentence. Discuss it with others.

   Area of a triangle = _____

   _____

   _____

3. Examine the results of Figures G–J. Propose a formula for the area of a parallelogram as an equation and as a word sentence. Discuss it with others.

   Area of a parallelogram = _____

   _____

   _____

Defining *Base* and *Height*

Study the figures below. Then write definitions for the words **base** and **height**.

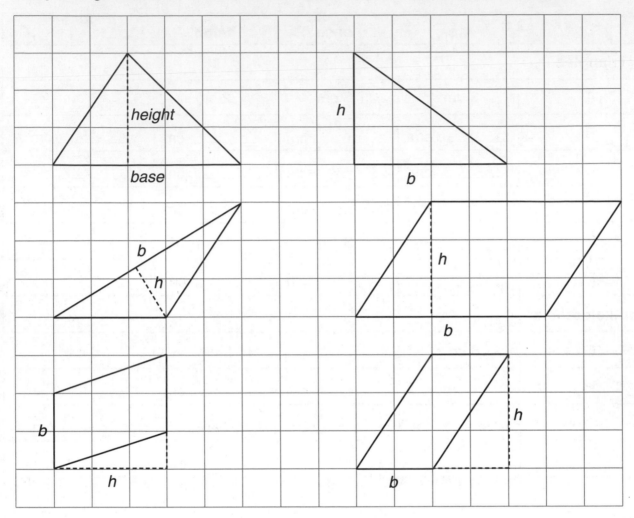

base:

_____

_____

height:

_____

_____

_____

Use with Lesson 9.6.

Math Boxes 9.6

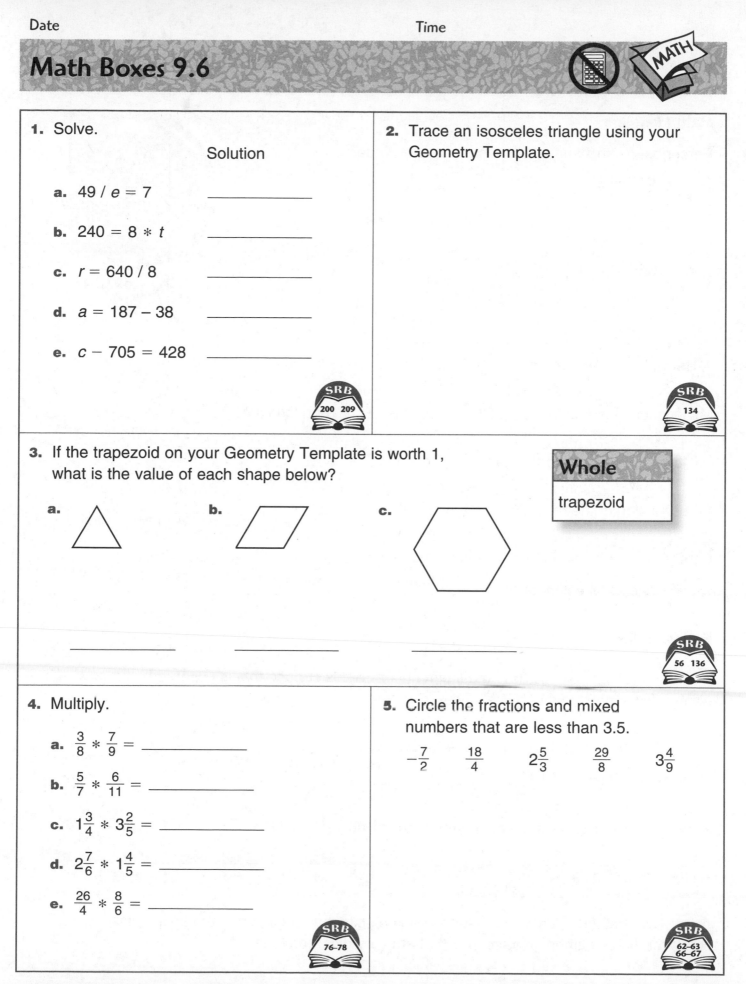

1. Solve.

Solution

a. $49 / e = 7$ _____

b. $240 = 8 * t$ _____

c. $r = 640 / 8$ _____

d. $a = 187 - 38$ _____

e. $c - 705 = 428$ _____

SRB
200 209

2. Trace an isosceles triangle using your Geometry Template.

SRB
134

3. If the trapezoid on your Geometry Template is worth 1, what is the value of each shape below?

a.

b.

c.

Whole

trapezoid

_____   _____   _____

SRB
56 136

4. Multiply.

a. $\frac{3}{8} * \frac{7}{9} =$ _____

b. $\frac{5}{7} * \frac{6}{11} =$ _____

c. $1\frac{3}{4} * 3\frac{2}{5} =$ _____

d. $2\frac{7}{6} * 1\frac{4}{5} =$ _____

e. $\frac{26}{4} * \frac{8}{6} =$ _____

SRB
76–78

5. Circle the fractions and mixed numbers that are less than 3.5.

$-\frac{7}{2}$ $\frac{18}{4}$ $2\frac{5}{3}$ $\frac{29}{8}$ $3\frac{4}{9}$

SRB
62–63
66–67

Earth's Water Surface

Math Message

Percent of Earth's surface that is covered by water:

    My estimate: _____

A Sampling Experiment

My location is at latitude _____ and longitude _____.

My location is on   land   water.   (Circle one.)

What fraction of the class has a water location? _____

Percent of Earth's surface that is covered by water:

    My class's estimate: _____

Follow-Up

Your teacher can tell you the actual percent of Earth's surface that is covered by water, or you can look it up in a reference book.

Percent of Earth's surface that is covered by water:

    Actual figure: _____

How does your class's estimate compare to the actual figure?

_____

_____

Note: This method of sampling usually gives results that are close to the actual value. However, it sometimes gives results that are very different.

Estimation Challenge: Area

What is the ground area of your school? In other words, what area of land is taken up by the ground floor?

Work alone or with a partner to come up with an estimation plan. How can you estimate the ground area of your school without measuring it with a tape measure? Discuss your ideas with your classmates.

My estimation plan:

_____

_____

_____

_____

_____

My best estimate:

_____

_____

_____

_____

_____

How accurate is your estimate? What range of areas might the actual area fall in?

_____

_____

_____

_____

_____

_____

The Four-4s Problem

Using only four 4s and any operation on your calculator, create expressions for values from 1 through 100. Do not use any other numbers except for the ones listed in the rules below. You do not need to find an expression for every number. Some are quite difficult. Try to find as many as you can today, but keep working when you have free time. The rules are listed below:

- You must use four 4s in every expression.
- You can use two 4s to create 44 or $\frac{4}{4}$. You can use three 4s to create 444.
- You may use 4^0. ($4^0 = 1$)
- You may use $\sqrt{4}$. ($\sqrt{4} = 2$)
- You may use 4! (four factorial). ($4! = 4 * 3 * 2 * 1 = 24$)

Use parentheses as needed so that it is very clear what is to be done and in what order. Examples of expressions for some numbers are shown below.

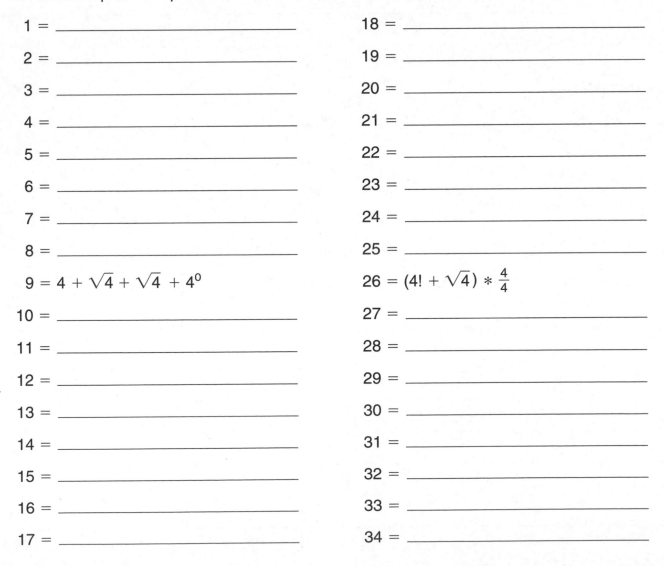

1 = _____

2 = _____

3 = _____

4 = _____

5 = _____

6 = _____

7 = _____

8 = _____

9 = $4 + \sqrt{4} + \sqrt{4} + 4^0$

10 = _____

11 = _____

12 = _____

13 = _____

14 = _____

15 = _____

16 = _____

17 = _____

18 = _____

19 = _____

20 = _____

21 = _____

22 = _____

23 = _____

24 = _____

25 = _____

26 = $(4! + \sqrt{4}) * \frac{4}{4}$

27 = _____

28 = _____

29 = _____

30 = _____

31 = _____

32 = _____

33 = _____

34 = _____

                                               Use with Lesson 9.7.

The Four-4s Problem (cont.)

35 = _____

36 = _____

37 = _____

38 = _____

39 = _____

40 = _____

41 = _____

42 = _____

43 = _____

44 = _____

45 = _____

46 = _____

47 = _____

48 = _____

49 = _____

50 = _____

51 = _____

52 = _____

53 = _____

54 = _____

55 = _____

56 = _____

57 = _____

58 = $[(\sqrt{4} * (4! + 4)] + \sqrt{4}$

59 = _____

60 = _____

61 = _____

62 = _____

63 = _____

64 = _____

65 = _____

66 = _____

67 = _____

68 = _____

69 = _____

70 = _____

71 = _____

72 = _____

73 = _____

74 = _____

75 = _____

76 = _____

77 = _____

78 = _____

79 = _____

80 = _____

81 = _____

82 = _____

83 = _____

84 = _____

The Four-4s Problem (cont.)

85 = _____   94 = _____

86 = _____   95 = _____

87 = _____   96 = _____

88 = _____   97 = _____

89 = _____   98 = _____

90 = _____   99 = _____

91 = _____   100 = _____

92 = _____

93 = _____

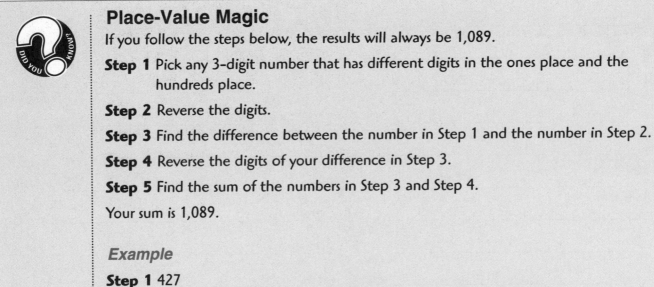

Place-Value Magic

If you follow the steps below, the results will always be 1,089.

Step 1 Pick any 3-digit number that has different digits in the ones place and the hundreds place.

Step 2 Reverse the digits.

Step 3 Find the difference between the number in Step 1 and the number in Step 2.

Step 4 Reverse the digits of your difference in Step 3.

Step 5 Find the sum of the numbers in Step 3 and Step 4.

Your sum is 1,089.

Example

Step 1 427

Step 2 724

Step 3 724 − 427 = 297

Step 4 792

Step 5 792 + 297 = 1,089

Can you figure out why this works?

Math Boxes 9.7

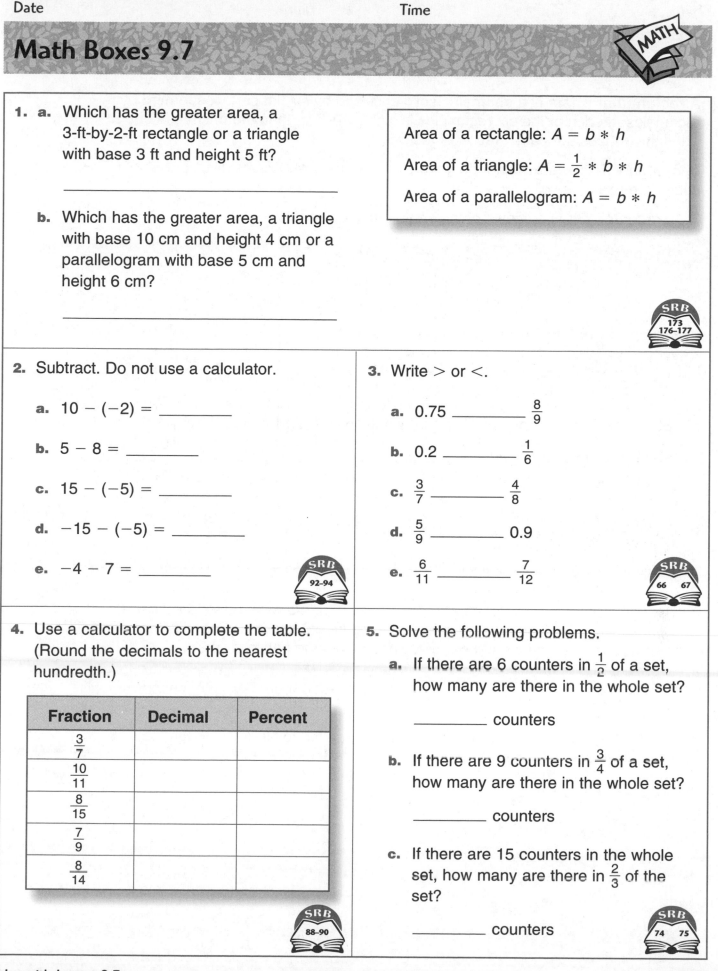

1. **a.** Which has the greater area, a 3-ft-by-2-ft rectangle or a triangle with base 3 ft and height 5 ft?

_____

b. Which has the greater area, a triangle with base 10 cm and height 4 cm or a parallelogram with base 5 cm and height 6 cm?

_____

Area of a rectangle: $A = b * h$

Area of a triangle: $A = \frac{1}{2} * b * h$

Area of a parallelogram: $A = b * h$

SRB
173
176–177

2. Subtract. Do not use a calculator.

a. $10 - (-2) =$ _____

b. $5 - 8 =$ _____

c. $15 - (-5) =$ _____

d. $-15 - (-5) =$ _____

e. $-4 - 7 =$ _____

SRB
92–94

3. Write > or <.

a. 0.75 _____ $\frac{8}{9}$

b. 0.2 _____ $\frac{1}{6}$

c. $\frac{3}{7}$ _____ $\frac{4}{8}$

d. $\frac{5}{9}$ _____ 0.9

e. $\frac{6}{11}$ _____ $\frac{7}{12}$

SRB
66   67

4. Use a calculator to complete the table. (Round the decimals to the nearest hundredth.)

Fraction	Decimal	Percent
$\frac{3}{7}$		
$\frac{10}{11}$		
$\frac{8}{15}$		
$\frac{7}{9}$		
$\frac{8}{14}$		

SRB
88–90

5. Solve the following problems.

a. If there are 6 counters in $\frac{1}{2}$ of a set, how many are there in the whole set?

_____ counters

b. If there are 9 counters in $\frac{3}{4}$ of a set, how many are there in the whole set?

_____ counters

c. If there are 15 counters in the whole set, how many are there in $\frac{2}{3}$ of the set?

_____ counters

SRB
74   75

Rectangular Prisms

A **rectangular prism** is a geometric solid enclosed by six flat surfaces formed by rectangles. If each of the six rectangles is also a square, then the prism is a **cube.** The flat surfaces are called **faces** of the prism.

Bricks, paperback books, and most boxes are rectangular prisms. Dice and sugar cubes are examples of cubes.

Here are three different views of the same rectangular prism.

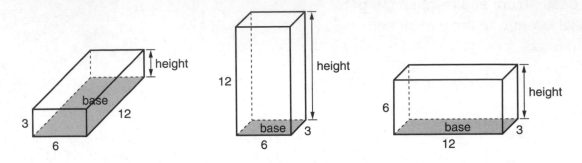

1. Study the figures above. Write your own definitions for **base** and **height.**

   Base of a rectangular prism: _____

   _____

   Height of a rectangular prism: _____

   _____

Examine the patterns on Activity Sheet 7. These patterns will be used to construct open boxes—boxes that have no tops. Try to figure out how many centimeter cubes are needed to fill each box to the top. Do not cut out the patterns yet.

2. I think that _____ centimeter cubes are needed to fill Box A to the top.

3. I think that _____ centimeter cubes are needed to fill Box B to the top.

Volumes of Rectangular Prisms

Write the formula for the volume of a rectangular prism.

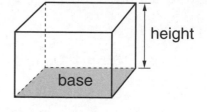

_____

B is the area of the base.

h is the height from that base.

V is the volume of the prism.

Find the volume of each rectangular prism below.

1.

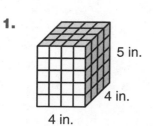

5 in.

4 in.

4 in.

V = _____

(unit)

2.

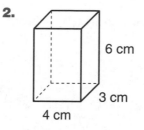

6 cm

3 cm

4 cm

V = _____

(unit)

3.

7 cm

7 cm

7 cm

V = _____

(unit)

4.

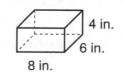

4 in.

6 in.

8 in.

V = _____

(unit)

5.

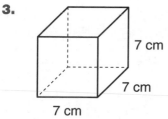

5 ft

3 ft

6 ft

V = _____

(unit)

6.

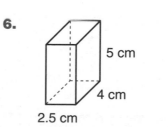

5 cm

4 cm

2.5 cm

V = _____

(unit)

A Mental Calculation Strategy

When you are multiplying mentally, it is sometimes helpful to double one factor and halve the other factor.

Example 1 45 * 12 = ?

Step 1 Double 45 and halve 12: 45 * 12 = 90 * 6.

Step 2 Multiply 90 and 6: 90 * 6 = 540.

Example 2 18 * 15 = ?

Step 1 Halve 18 and double 15: 18 * 15 = 9 * 30.

Step 2 Multiply 9 and 30: 9 * 30 = 270.

Example 3 75 * 28 = ?

Step 1 Double 75 to get 150 and halve 28 to get 14.

Step 2 Double again to get 300 and halve again to get 7.

Step 3 75 * 28 = 300 * 7 = 2,100.

Use the doubling and halving strategy to calculate mentally. Solve the problems below.

1. 35 * 14 = _____

New number sentence:

_____

2. 16 * 25 = _____

New number sentence:

_____

3. 18 * 35 = _____

New number sentence:

_____

4. 15 * 44 = _____

New number sentence:

_____

5. 14 * 55 = _____

New number sentence:

_____

6. 75 * 24 = _____

New number sentence:

_____

New number sentence:

_____

Use with Lesson 9.8.

Math Boxes 9.8

1. Plot and label the ordered number pairs on the grid.

   E: (−2,5)

   F: (3,4)

   G: (−2,−4)

   H: (−1,0)

   I: (5,−1)

   J: (4,4)

2. Mr. Carroll's class collected autumn leaves and sorted them by color. The class had 24 yellow leaves, 17 green leaves, 37 red leaves, 8 orange leaves, and 14 brown leaves. Make a circle graph of the data.

   _____
   (title)

3. Add the fractions.

   a. $\frac{1}{4} + \frac{1}{6} =$ _____

   b. $\frac{3}{3} + \frac{1}{12} =$ _____

   c. $\frac{2}{12} + \frac{2}{4} =$ _____

   d. $\frac{5}{12} + \frac{2}{3} =$ _____

4. Write a number story for 185 / 6 and solve it.

   _____

   _____

   _____

   Answer: _____

Volume of Prisms

The volume V of any prism can be found with the formula $V = B * h$, where B is the area of the base of the prism, and h is the height of the prism for that base.

Find the volume of each prism.

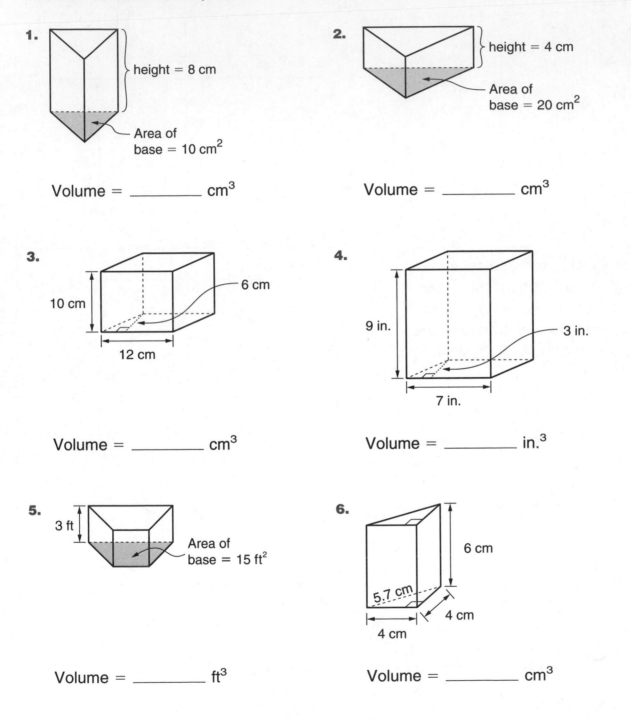

1.

height = 8 cm

Area of base = 10 cm^2

Volume = _____ cm^3

2.

height = 4 cm

Area of base = 20 cm^2

Volume = _____ cm^3

3.

10 cm

6 cm

12 cm

Volume = _____ cm^3

4.

9 in.

3 in.

7 in.

Volume = _____ in.^{3}

5.

3 ft

Area of base = 15 ft^2

Volume = _____ ft^3

6.

6 cm

5.7 cm

4 cm

4 cm

Volume = _____ cm^3

Use with Lesson 9.9.

Volume of Prisms (cont.)

7.

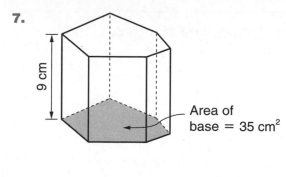

Area of base = 35 cm²

Volume = _____ cm³

8.

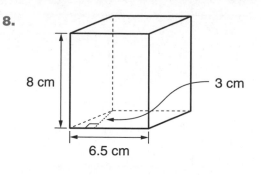

8 cm

3 cm

6.5 cm

Volume = _____ cm³

9.

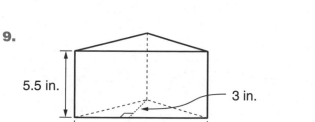

5.5 in.

3 in.

11 in.

Volume = _____ in.³

10.

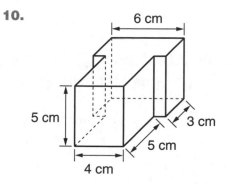

6 cm

5 cm

3 cm

5 cm

4 cm

Volume = _____ cm³

11.

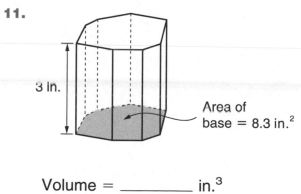

3 in.

Area of base = 8.3 in.²

Volume = _____ in.³

12.

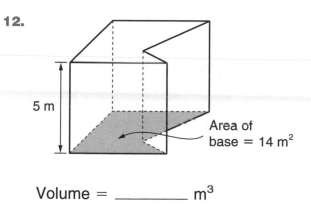

5 m

Area of base = 14 m²

Volume = _____ m³

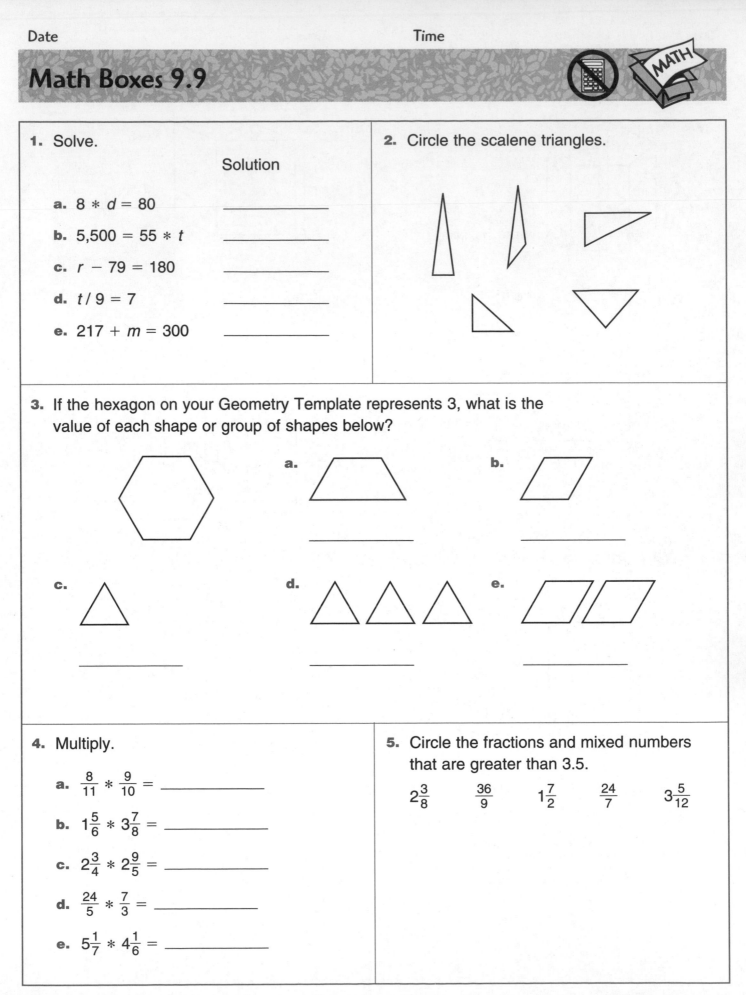

1. Solve.

Solution

a. $8 * d = 80$ _____

b. $5,500 = 55 * t$ _____

c. $r - 79 = 180$ _____

d. $t / 9 = 7$ _____

e. $217 + m = 300$ _____

2. Circle the scalene triangles.

3. If the hexagon on your Geometry Template represents 3, what is the value of each shape or group of shapes below?

a. _____

b. _____

c. _____

d. _____

e. _____

4. Multiply.

a. $\dfrac{8}{11} * \dfrac{9}{10} =$ _____

b. $1\dfrac{5}{6} * 3\dfrac{7}{8} =$ _____

c. $2\dfrac{3}{4} * 2\dfrac{9}{5} =$ _____

d. $\dfrac{24}{5} * \dfrac{7}{3} =$ _____

e. $5\dfrac{1}{7} * 4\dfrac{1}{6} =$ _____

5. Circle the fractions and mixed numbers that are greater than 3.5.

$2\dfrac{3}{8}$ $\quad$ $\dfrac{36}{9}$ $\quad$ $1\dfrac{7}{2}$ $\quad$ $\dfrac{24}{7}$ $\quad$ $3\dfrac{5}{12}$

Units of Volume and Capacity

In the metric system, units of length, volume, capacity, and weight are related.

- The **cubic centimeter (cm³)** is a metric unit of volume.

- The **liter (L)** and **milliliter (mL)** are units of capacity.

1. Complete.

    a. 1 liter (L) = _____ milliliters (mL).

    b. There are _____ cubic centimeters (cm³) in 1 liter.

    c. So 1 cm³ = _____ mL.

2. The cube in the diagram has sides 5 cm long.

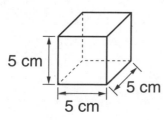

5 cm 5 cm 5 cm

    a. What is the volume of the cube?

    _____ cm³

    b. If the cube were filled with water, how many milliliters would it hold?

    _____ mL

3. a. What is the volume of the rectangular prism in the drawing?

    _____ cm³

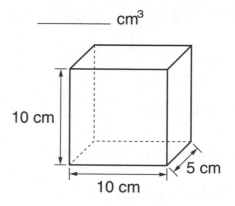

10 cm 10 cm 5 cm

    b. If the prism were filled with water, how many milliliters would it hold?

    _____ mL

    c. That is what fraction of a liter?

    _____ L

Complete.

4. 2 L = _____ mL

5. 350 cm³ = _____ mL

6. 1,500 mL = _____ L

7. One liter of water weighs about 1 kilogram (kg).

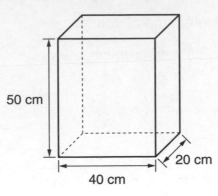

50 cm

20 cm

40 cm

If the tank in the diagram above is filled with water, about how much will the water

weigh? _____ kg

In the U.S. customary system, units of length and capacity are not closely related.
Larger units of capacity are multiples of smaller units.

- 1 cup (c) = 8 fluid ounces (fl oz)

- 1 pint (pt) = 2 cups (c)

- 1 quart (qt) = 2 pints (pt)

- 1 gallon (gal) = 4 quarts (qt)

8. **a.** 1 gallon = _____ quarts

   b. 1 gallon = _____ pints

9. **a.** 2 quarts = _____ pints

   b. 2 quarts = _____ fluid ounces

10. Sometimes it is helpful to know that 1 liter is a little more than 1 quart. In the
United States, gasoline is sold by the gallon. If you travel in Canada or Mexico,
you will find that gasoline is sold by the liter. Is 1 gallon of gasoline more or less
than 4 liters of gasoline?

_____

Use with Lesson 9.10.

Open Boxes

What are the dimensions of an open box—having the greatest possible volume—that can be made out of a single sheet of centimeter grid paper?

1. Use centimeter grid paper to experiment until you discover a pattern. Record your results in the table below.

height of box	length of base	width of base	Volume of box
1 cm	20 cm	14 cm	
2 cm			
3 cm			

2. What are the dimensions of the box with the greatest volume?

height of box = _____ cm   length of base = _____ cm

width of base = _____ cm   Volume of box = _____ cm³

1. Circle the figure below that has the same area as Figure A.

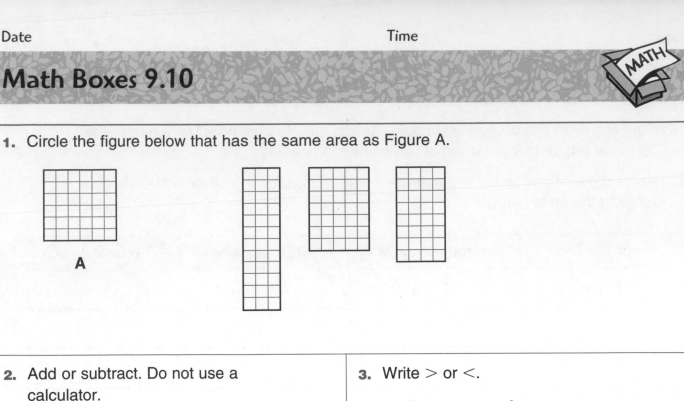

   A

2. Add or subtract. Do not use a calculator.

   a. $-22 + 12 =$ _____

   b. $18 - (-4) =$ _____

   c. $-15 - (-8) =$ _____

   d. $-4 + (-17) =$ _____

   e. $-6 - (-28) =$ _____

3. Write > or <.

   a. $\frac{7}{8}$ _____ $\frac{9}{10}$

   b. $\frac{4}{5}$ _____ 0.89

   c. $\frac{2}{3}$ _____ $\frac{5}{8}$

   d. 0.37 _____ $\frac{2}{5}$

   e. $\frac{9}{6}$ _____ 1.05

4. Use a calculator to complete the table. (Round the decimals to the nearest hundredth.)

Fraction	Decimal	Percent
$\frac{11}{12}$		
$\frac{5}{7}$		
$\frac{14}{15}$		
$\frac{5}{6}$		
$\frac{2}{9}$		

5. Solve.

   a. $\frac{1}{3}$ of 27 = _____

   b. $\frac{1}{8}$ of 40 = _____

   c. $\frac{1}{5}$ of 100 = _____

   d. $\frac{2}{5}$ of 100 = _____

   e. $\frac{1}{4}$ of 60 = _____

Time to Reflect

1. How would you explain to someone the difference between area and volume?

_____

_____

_____

_____

2. Describe at least three situations where you would want to find the area of something.

_____

_____

_____

_____

3. Describe at least one situation where you would want to find the volume of something.

_____

_____

_____

_____

4. Explain the difference between plotting the point (3,5) and plotting the point (5,3).

_____

_____

_____

_____

Math Boxes 9.11

1. Plot and label the ordered number pairs on the grid.

   U: (−6,0)

   V: (2,−5)

   W: (−4,−3)

   X: (4,3)

   Y: (0,−2)

   Z: (−3,5)

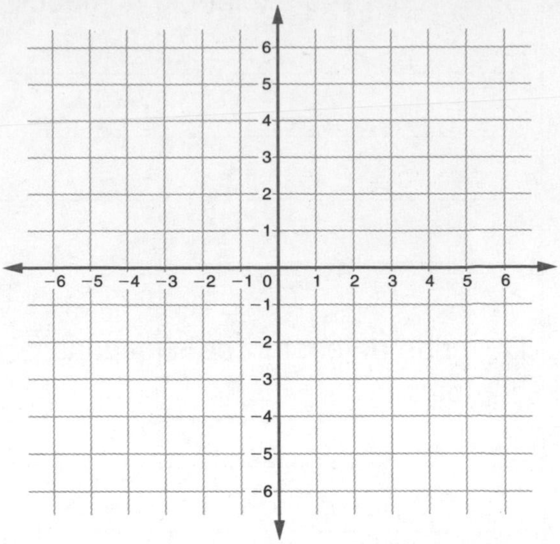

2. Complete the "What's My Rule?" table and state the rule.

   Rule: _____

in	out
4	9
7	15
11	23
	19
6	

3. If the radius of a circle is 2.5 inches, what is its diameter?

   _____
   (unit)

   Explain. _____

   _____

   _____

4. Explain how you could find the area of the rectangle below.

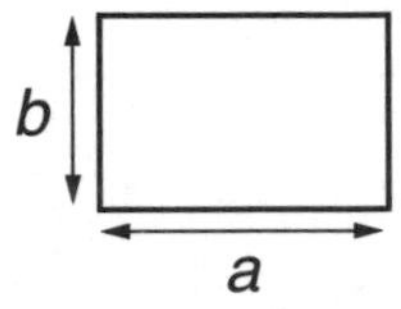

   b │ ⬜ a

   _____

   _____

   _____

   _____

5. Solve.

   Solution

   a. $\frac{3}{8} = \frac{a}{40}$ _____

   b. $-80 + c = 100$ _____

   c. $m * 25 = 400$ _____

   d. $s - 110 = -20$ _____

   e. $144 / z = 12$ _____

Use with Lesson 9.11.

Math Boxes 10.1

1. Measure the base and height of the triangle below to the nearest centimeter.

The base is about _____ cm.

The height is about _____ cm.

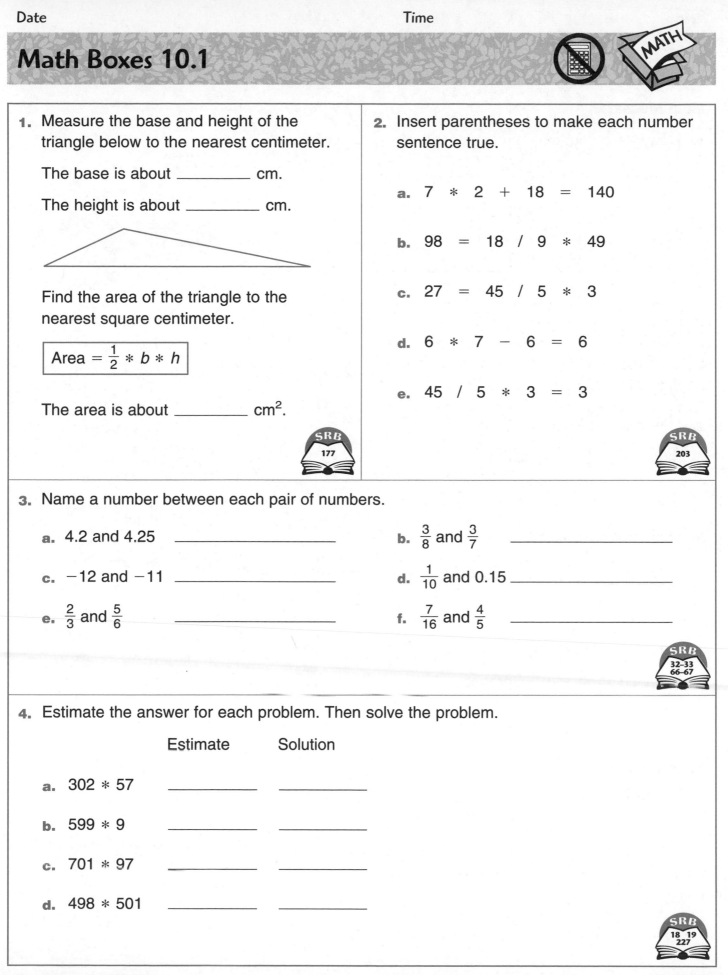

Find the area of the triangle to the nearest square centimeter.

$$\text{Area} = \frac{1}{2} * b * h$$

The area is about _____ cm².

SRB 177

2. Insert parentheses to make each number sentence true.

a. $7 * 2 + 18 = 140$

b. $98 = 18 / 9 * 49$

c. $27 = 45 / 5 * 3$

d. $6 * 7 - 6 = 6$

e. $45 / 5 * 3 = 3$

SRB 203

3. Name a number between each pair of numbers.

a. 4.2 and 4.25 _____

b. $\frac{3}{8}$ and $\frac{3}{7}$ _____

c. -12 and -11 _____

d. $\frac{1}{10}$ and 0.15 _____

e. $\frac{2}{3}$ and $\frac{5}{6}$ _____

f. $\frac{7}{16}$ and $\frac{4}{5}$ _____

SRB 32–33 66–67

4. Estimate the answer for each problem. Then solve the problem.

	Estimate	Solution
a. $302 * 57$	_____	_____
b. $599 * 9$	_____	_____
c. $701 * 97$	_____	_____
d. $498 * 501$	_____	_____

SRB 18 19 227

Pan-Balance Problems

Math Message

1. How would you use a pan balance to weigh an object?

   _____

   _____

   _____

   _____

Solve these pan-balance problems. In each figure, the two pans are in perfect balance.

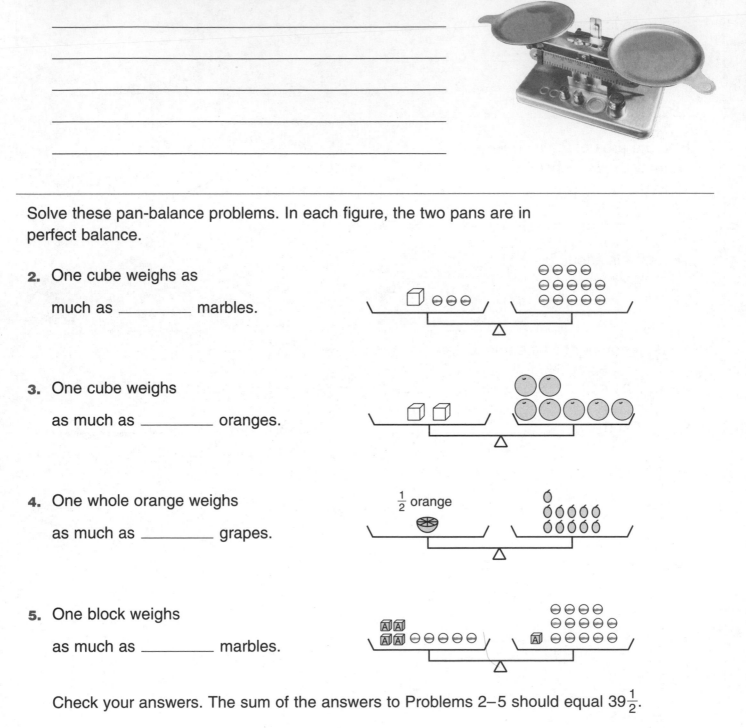

2. One cube weighs as

   much as _____ marbles.

3. One cube weighs

   as much as _____ oranges.

4. One whole orange weighs

   as much as _____ grapes.

   $\frac{1}{2}$ orange

5. One block weighs

   as much as _____ marbles.

Check your answers. The sum of the answers to Problems 2–5 should equal $39\frac{1}{2}$.

Pan-Balance Problems (cont.)

6. One ☐ weighs

as much as _____ △s.

7. One ☐ weighs

as much as _____ marbles.

8. One *x* weighs

as much as _____ balls.

9. One *u* weighs

as much as _____ *v*s.

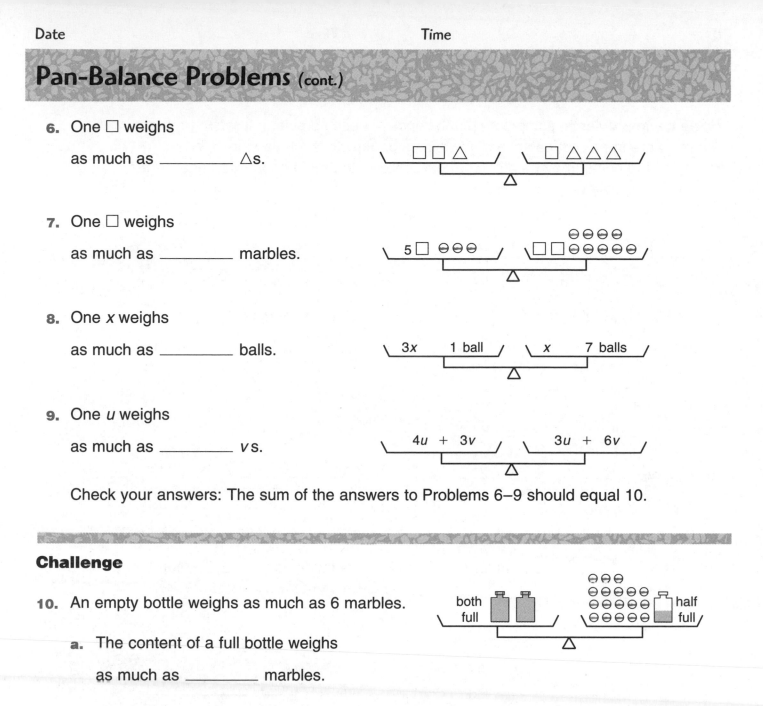

Check your answers: The sum of the answers to Problems 6–9 should equal 10.

Challenge

10. An empty bottle weighs as much as 6 marbles.

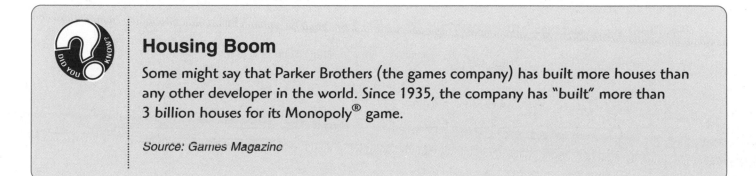

a. The content of a full bottle weighs

as much as _____ marbles.

b. A full bottle weighs as much as _____ marbles.

Housing Boom

Some might say that Parker Brothers (the games company) has built more houses than any other developer in the world. Since 1935, the company has "built" more than 3 billion houses for its Monopoly® game.

Source: Games Magazine

More Pan-Balance Problems

Solve these problems, using both pan balances in each problem. In each problem, the pans are in perfect balance. The weights of objects, such as blocks, balls, marbles, and coins, may be different from problem to problem, but are consistent within each problem.

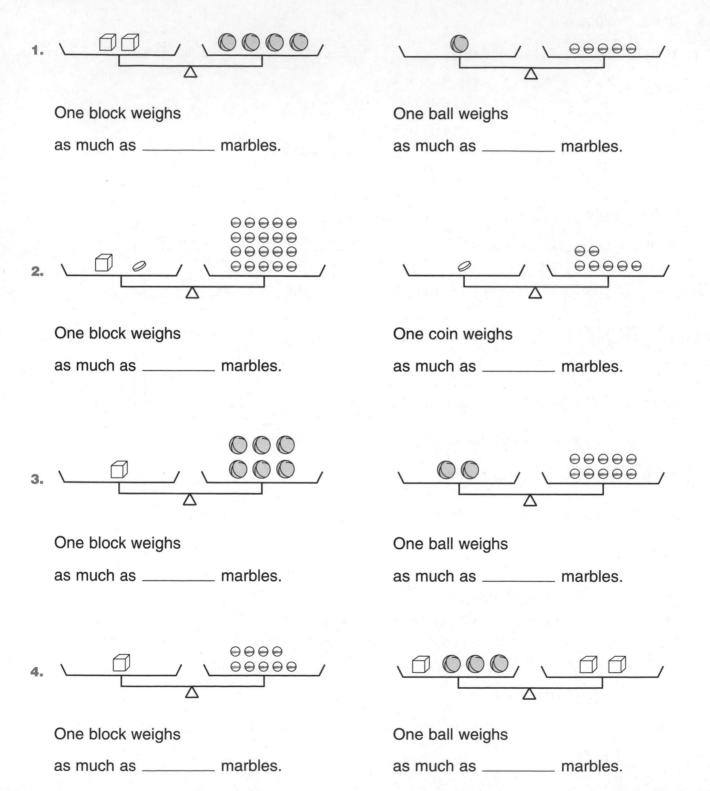

1.

One block weighs

as much as _____ marbles.

One ball weighs

as much as _____ marbles.

2.

One block weighs

as much as _____ marbles.

One coin weighs

as much as _____ marbles.

3.

One block weighs

as much as _____ marbles.

One ball weighs

as much as _____ marbles.

4.

One block weighs

as much as _____ marbles.

One ball weighs

as much as _____ marbles.

Use with Lesson 10.2.

More Pan-Balance Problems (cont.)

5.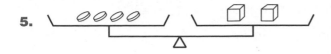

One coin weighs

as much as _____ clips.

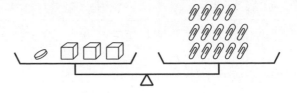

One block weighs

as much as _____ clips.

6.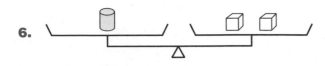

One can weighs

as much as _____ blocks.

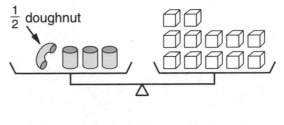

One doughnut weighs

as much as _____ blocks.

7.

One □ weighs

as much as _____ marbles.

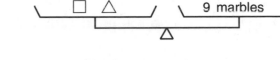

One △ weighs

as much as _____ marbles.

8.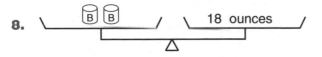

Each can weighs *B* ounces.

B = _____ ounces

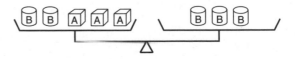

Each cube weighs *A* ounces.

A = _____ ounces

More Pan-Balance Problems (cont.)

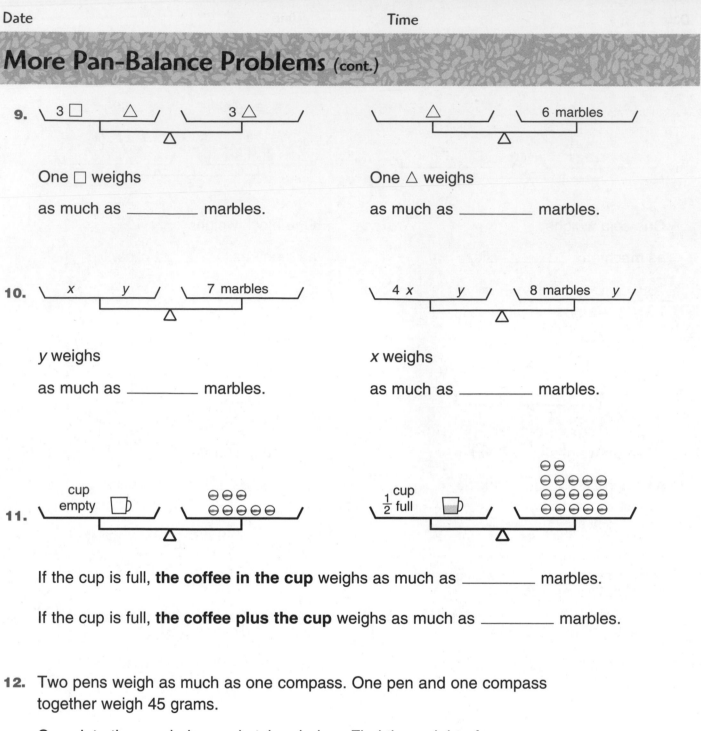

9.

One □ weighs

as much as _____ marbles.

One △ weighs

as much as _____ marbles.

10.

y weighs

as much as _____ marbles.

x weighs

as much as _____ marbles.

11.

If the cup is full, **the coffee in the cup** weighs as much as _____ marbles.

If the cup is full, **the coffee plus the cup** weighs as much as _____ marbles.

12. Two pens weigh as much as one compass. One pen and one compass together weigh 45 grams.

Complete the pan-balance sketches below. Find the weight of one pen and of one compass.

a.

One pen weighs _____ grams.

b.

One compass weighs _____ grams.

Use with Lesson 10.2.

Represent Number Stories

Circle each expression that correctly represents the information in the story.
(There may be more than one answer.)

1. Melissa baked 5 trays with 12 cookies each. She sold 3 trays of 12 cookies.

Number of cookies sold:

$(5 - 3) * 12$ $3 * 12$ $5 * 12$

2. Jonas was stocking up on soda for the family. He bought 8 six-packs of soda. His mom bought 3 more six-packs.

Total number of cans of soda:

$(8 * 3) + 6$ $(8 * 3) + (8 * 6)$ $6 * (8 + 3)$

3. Jenny bought 6 envelopes for 14 cents each and 6 stamps for 34 cents each.

Amount of money spent:

$(6 + 6) * (14 + 34)$ $(6 * 14) + (6 * 34)$ $6 * (14 + 34)$

4. Monty had 8 packages of 12 pencils when the year started. He used 3 packages of 12 during the school year.

Number of pencils left:

$(8 - 3) * 12$ $(8 * 12) - (3 * 12)$ $(8 + 3) * 12$

5. Make up one of your own.

_____

_____

_____

Your number expression: _____

Math Boxes 10.2

1. Draw a figure on the grid that has a perimeter of 32 units.

1 unit

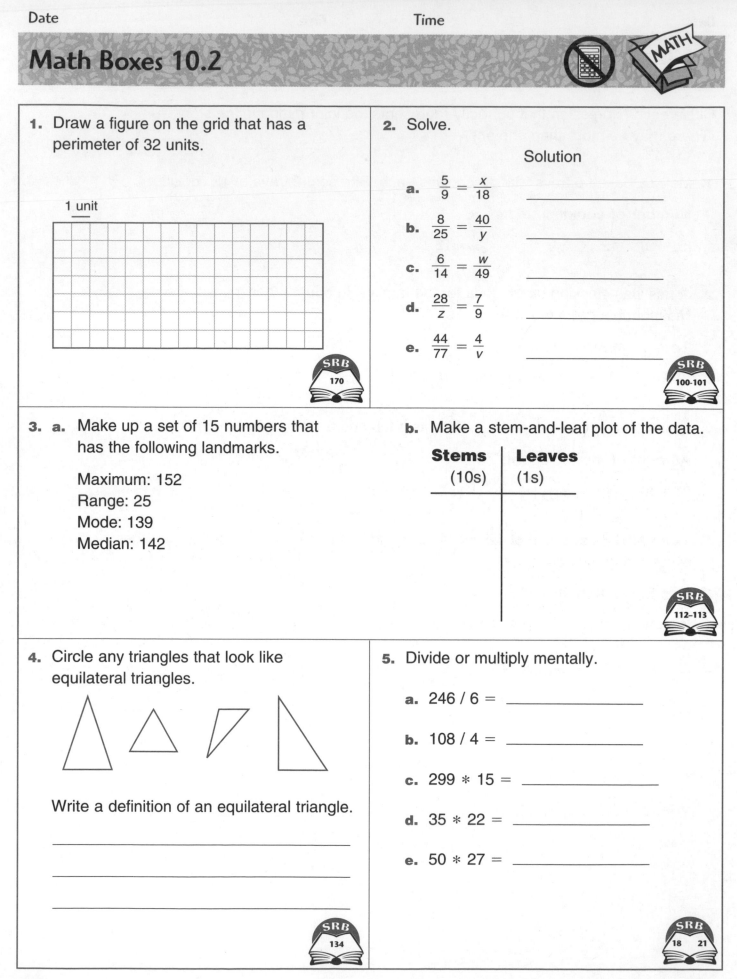

SRB 170

2. Solve.

Solution

a. $\dfrac{5}{9} = \dfrac{x}{18}$ _____

b. $\dfrac{8}{25} = \dfrac{40}{y}$ _____

c. $\dfrac{6}{14} = \dfrac{w}{49}$ _____

d. $\dfrac{28}{z} = \dfrac{7}{9}$ _____

e. $\dfrac{44}{77} = \dfrac{4}{v}$ _____

SRB 100-101

3. a. Make up a set of 15 numbers that has the following landmarks.

Maximum: 152
Range: 25
Mode: 139
Median: 142

b. Make a stem-and-leaf plot of the data.

Stems (10s)	Leaves (1s)

SRB 112–113

4. Circle any triangles that look like equilateral triangles.

Write a definition of an equilateral triangle.

_____

_____

_____

SRB 134

5. Divide or multiply mentally.

a. 246 / 6 = _____

b. 108 / 4 = _____

c. 299 * 15 = _____

d. 35 * 22 = _____

e. 50 * 27 = _____

SRB 18 21

Math Boxes 10.3

1. Find the area and perimeter of the rectangle.

7 cm

$3\frac{1}{2}$ cm

Area = _____
        (unit)

Perimeter = _____
            (unit)

2. Insert parentheses to make each number sentence true.

a. 6 + 8 * 10 = 140

b. 21 = 42 / 6 − 4

c. 7 * 7 + 2 = 63

d. 3 * 15 − 3 = 36

e. 42 / 6 − 4 = 3

3. Name a number between each pair of numbers.

a. −1.30 and −1.20 _____

b. 8.05 and 8.10 _____

c. −0.26 and −0.25 _____

d. $\frac{1}{3}$ and $\frac{7}{8}$ _____

e. $\frac{1}{4}$ and 0.3 _____

f. 0.2 and $\frac{2}{9}$ _____

4. Estimate the answer for each problem. Then solve the problem.

Estimate   Solution

a. 60.3 * 71 _____

b. 29 * 0.8 _____   _____

c. 48 * 2.02 _____   _____

d. 2.2 * 550 _____   _____

Algebraic Expressions

Complete each statement below with an algebraic expression, using the suggested variable. The first problem has been done for you.

1. If Beth's allowance is $2.50 more than Ann's, then Beth's allowance is

   ___*D + $2.50*___.

Ann's allowance is *D* dollars.　　　Beth

2. If John gets a raise of $5 per week, then his salary is

   $ _____.

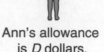

John's salary is *S* dollars per week.

3. If Ali's grandfather is 50 years older than Ali, then Ali is

   _____ years old.

Ali's grandfather is *G* years old.　　　Ali

4. Seven baskets of potatoes weigh

   _____ pounds.

A basket of potatoes weighs *P* pounds.

Algebraic Expressions (cont.)

5. If a submarine dives 150 feet, then it will be traveling at a depth

of _____ feet.

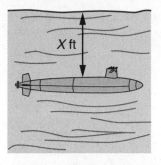

A submarine is traveling at a depth of *X* feet.

6. The floor is divided up for gym classes into 5 equal-sized areas. Each class has a playing area of

_____ ft².

The gym floor has an area of *A* square feet.

7. The charge for a book that is *D* days overdue is

_____ cents.

A library charges 10 cents for each overdue book. It adds an additional charge of 5 cents per day for each overdue book.

8. If Kevin spends $\frac{2}{3}$ of his allowance on a book, then he has

_____ dollars left.

Kevin's allowance is *X* dollars.

"What's My Rule?"

1. a. State in words the rule for the "What's My Rule?" table at the right.

X	Y
5	1
4	0
−1	−5
1	−3
2	−2

_____

_____

_____

_____

b. Circle the number sentence that describes the rule.

$Y = X / 5$ $Y = X − 4$ $Y = 4 − X$

2. a. State in words the rule for the "What's My Rule?" table at the right.

Q	Z
1	3
3	5
−4	−2
−3	−1
−2.5	−0.5

_____

_____

_____

_____

b. Circle the number sentence that describes the rule.

$Z = Q + 2$ $Z = 2 * Q$ $Z = \frac{1}{2}Q * 1$

3. a. State in words the rule for the "What's My Rule?" table at the right.

g	t
$\frac{1}{2}$	2
0	0
2.5	10
$\frac{1}{4}$	1
5	20

_____

_____

_____

_____

b. Circle the number sentence that describes the rule.

$g = 2 * t$ $t = 2 * g$ $t = 4 * g$

Multiplication and Division Practice

Solve. Show your work in the space below the problems.

1. $384 * 1.5 =$ _____

2. $50.3 * 89 =$ _____

3. $824 * 75 =$ _____

4. $\dfrac{843}{7} =$ _____

5. $70.4 / 8 =$ _____

6. $1,435 / 0.5 =$ _____

Speed and Distance

Math Message

1. A plane travels at a **speed** of 480 miles per hour. At that rate, how many miles will it travel in 1 minute? Write a number model to show what you did to solve the problem.

   Number model: _____   Distance per minute: _____ miles

Rule for Distance Traveled

2. For an airplane flying at 8 miles per minute (480 mph), you can use the following rule to calculate the distance traveled for any number of minutes:

   > Distance traveled = 8 * number of minutes
   >   or
   > $d = 8 * t$

   where d stands for the distance traveled and t for the time of travel, in minutes. For example, after 1 minute, the plane will have traveled 8 miles (8 * 1). After 2 minutes, it will have traveled 16 miles (8 * 2).

3. Use the rule $d = 8 * t$ to complete the table at the right.

Time (min) (t)	Distance (mi) (8 * t)
1	8
2	16
3	
4	
5	
6	
7	
8	
9	
10	

Use with Lesson 10.4.

Speed and Distance (cont.)

4. Complete the graph using the data in the table on page 354. Then connect the dots.

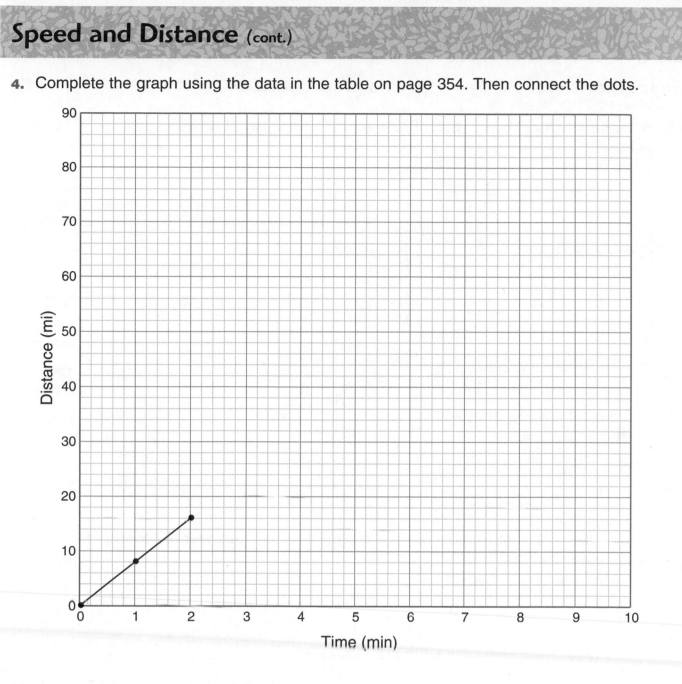

Use your graph to answer the following questions:

5. How far would the plane travel in $1\frac{1}{2}$ minutes?   _____
(unit)

6. How many miles would the plane travel in
5 minutes 24 seconds (5.4 minutes)?   _____
(unit)

7. How long would it take the plane to travel 60 miles?   _____
(unit)

Use with Lesson 10.4.

Representing Rates

Complete each table below. Then graph the data and connect the points.

1. a. Andy earns $8 per hour.   Rule: Earnings = $8 * number of hours worked

Time (hr) (h)	Earnings ($) (8 * h)
1	
2	
3	
	40
7	

b. Plot a point to show Andy's earnings for $5\frac{1}{2}$ hours. How much would he earn?

_____

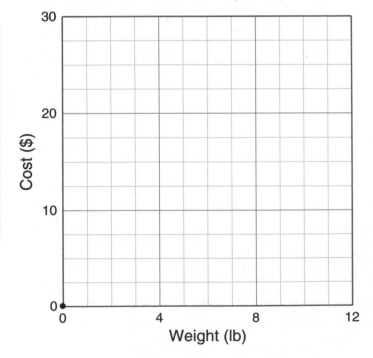

2. a. Red peppers cost $2.50 a pound.   Rule: Cost = $2.50 * number of pounds

Weight (lb) (w)	Cost ($) (2.50 * w)
1	
2	
3	
	15.00
12	

b. Plot a point to show the cost of 8 pounds. How much would 8 pounds of red peppers cost?

_____

Representing Rates (cont.)

3. **a.** Frank types an average of 45 words a minute.

Rule: Words typed = 45 * number of minutes

Time (min) (t)	Words (45 * t)
1	
2	
3	
	225
6	

b. Plot a point to show the number of words Frank types in 4 minutes. How many words is that?

_____

4. **a.** Joan's car uses 1 gallon of gasoline every 28 miles.

Rule: Distance = 28 * number of gallons

Gasoline (gal) (g)	Distance (mi) (28 * g)
1	
2	
3	
	140
$5\frac{1}{2}$	

b. Plot a point to show how far the car would travel on 1.4 gallons of gasoline. How many miles would it go?

_____

Area and Volume Review

Area and Volume Formulas

Area of rectangles and parallelograms

$A = b * h$, where A is the area, b is the length of the base, and h is the width or height

Area of triangles

$A = \frac{1}{2} * b * h$, where A is the area, b is the length of the base, and h is the height

Volume of prisms

$V = B * h$, where B is the area of the base and h is the height

Find the area of these figures.

1.

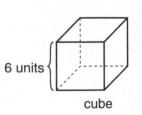

7 units

4 units

Area: _____

2.

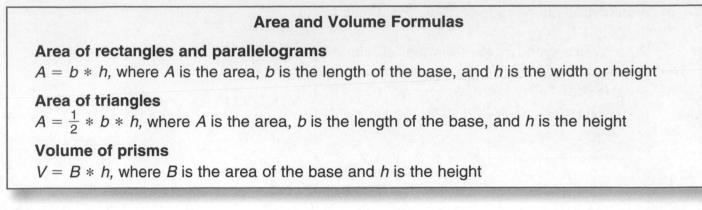

1 cm²

Area: _____

Find the volume of these solids.

3.

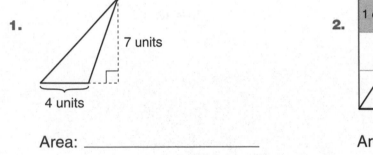

3 units

area of base
30 units²

Volume: _____

4.

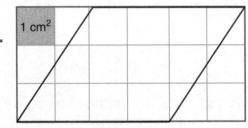

5 units

8 units

5 units

Volume: _____

5.

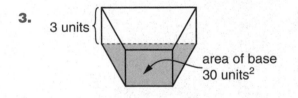

6 units

cube

Volume: _____

6.

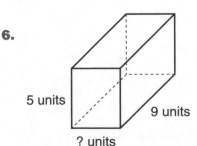

5 units

9 units

? units

The volume of this prism is 180 cubic units. What is the width of its base?

_____

Math Boxes 10.4

1. Draw a rectangle whose perimeter is the same as the perimeter of the rectangle shown but whose sides are not the same length as those shown.

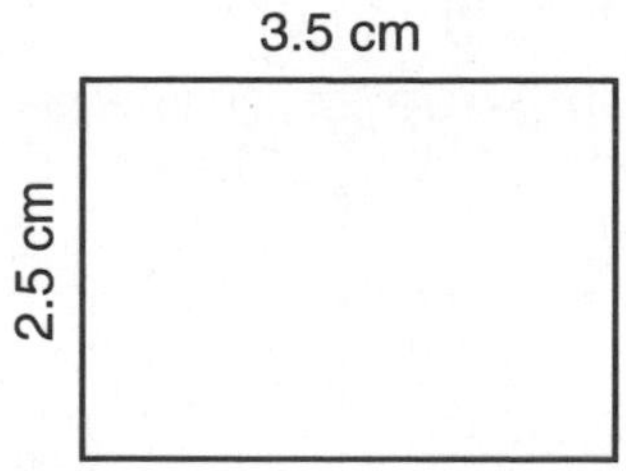

3.5 cm

2.5 cm

What is the area of the figure you drew?

_____

2. Solve.

Solution

a. $\dfrac{m}{10} = \dfrac{45}{50}$ _____

b. $\dfrac{56}{64} = \dfrac{7}{n}$ _____

c. $\dfrac{k}{48} = \dfrac{3}{8}$ _____

d. $\dfrac{4}{30} = \dfrac{12}{p}$ _____

e. $\dfrac{2}{18} = \dfrac{a}{180}$ _____

3. Make a stem-and-leaf plot of the following numbers:

120, 111, 137, 144, 121, 120, 95, 87, 120, 110, 135, 90, 86, 137, 144, 121, 120, 95, 87, 120, 110, 135, 90, 86, 143, 95, 141

Find the landmarks.

Mode: _____

Median: _____

Range: _____

Stems (10s)	Leaves (1s)

4. Draw an isosceles triangle.

Write a definition of an isosceles triangle.

_____

_____

_____

5. Divide or multiply mentally.

a. 495 / 5 = _____

b. 199 * 36 = _____

c. 63 * 500 = _____

d. 25 * 96 * 4 = _____

e. 843 / 3 = _____

Predicting When Old Faithful Will Erupt Next

Old Faithful Geyser in Yellowstone National Park is one of nature's most impressive sights. Yellowstone has 200 geysers and thousands of hot springs, mud pots, steam vents, and other "hot spots"—more than any other place on Earth. Old Faithful is not the largest or tallest geyser in Yellowstone, but it is the most dependable geyser. Using the length of time of an eruption, park rangers can predict when the next eruption will begin.

Old Faithful erupts at regular intervals that are **predictable.** If you time the length of one eruption, you can **predict** about how long you must wait until the next eruption. Use this formula:

Waiting time $= (10 * \text{(length of eruption)}) + 30$ minutes

$$W = (10 * E) + 30$$

or $W = 10E + 30$

All times are in minutes.

1. Use the formula to complete the table below.

Length of Eruption (min) (E)	Waiting Time to Next Eruption (min) ((10 * E) + 30)
2 min	50 min
3 min	_____ min
4 min	_____ min
5 min	_____ min
1 min	_____ min
$2\frac{1}{2}$ min	_____ min
3 min 15 sec	_____ min
_____ min	45 min

2. Graph the data from the table. One number pair has been plotted for you.

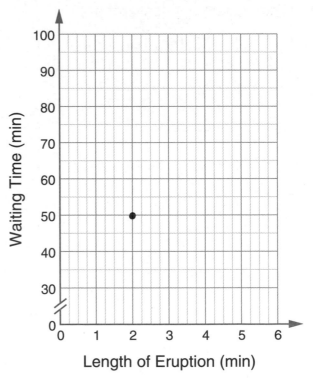

3. It's 8:30 A.M., and Old Faithful has just finished a 4-minute eruption. About when will it erupt next?

_____

4. The average time between eruptions of Old Faithful is about 75 minutes. So the average length of an eruption is about how many minutes?

_____

Use with Lesson 10.5.

More Pan-Balance Practice

Solve these pan-balance problems. In each figure, the two pans are in perfect balance.

1.

One orange weighs

as much as _____ triangle.

2.

One marble weighs

as much as _____ pencil.

3. 9 X   4 X

One doughnut weighs

as much as _____ Xs.

4. 17 P   3 S   6 P   36 S

One S weighs

as much as _____ P.

5.

One triangle weighs as much as _____ paper clips.

Explain how you found your answer. _____

_____

_____

_____

_____

_____

_____

6. 4 X   9 B   6 X   53 M   5 M   3 B   20 M   6 X

One X weighs as much as _____ M.

Math Boxes 10.5

1. Use your Geometry Template to trace three kinds of triangles in the space below. Under each triangle, write what kind of triangle it is.

SRB
134

2. Add or subtract.

a. $-7 + (-3) =$ _____

b. $5 - (-8) =$ _____

c. $-17 + 10 =$ _____

d. $-15 - 15 =$ _____

e. $3 + (-20) =$ _____

SRB
92–94

3. List all of the factors for 48.

_____

_____

SRB
10

4. Add or subtract.

a. $\frac{4}{5} + 1\frac{3}{8} =$ _____

b. $1\frac{2}{4} - \frac{4}{5} =$ _____

c. $6\frac{3}{7} - 3\frac{1}{3} =$ _____

d. _____ $= 4\frac{2}{9} + \frac{23}{6}$

SRB
70–72

5. The table shows how Robert spent his allowance for the month of April. Complete the table and make a circle graph of the data.

Type of Expense	Amount Spent	Percent of Allowance
Snacks	$2.50	
Movie	$5.50	
Gum	$0.50	
Baseball Cards	$1.50	
Total		

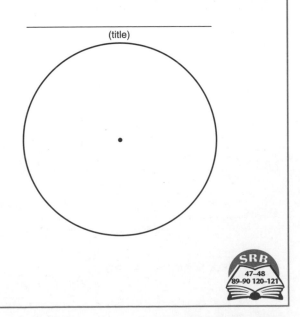

_____
(title)

SRB
47–48
89–90 120–121

Use with Lesson 10.5.

Math Boxes 10.6

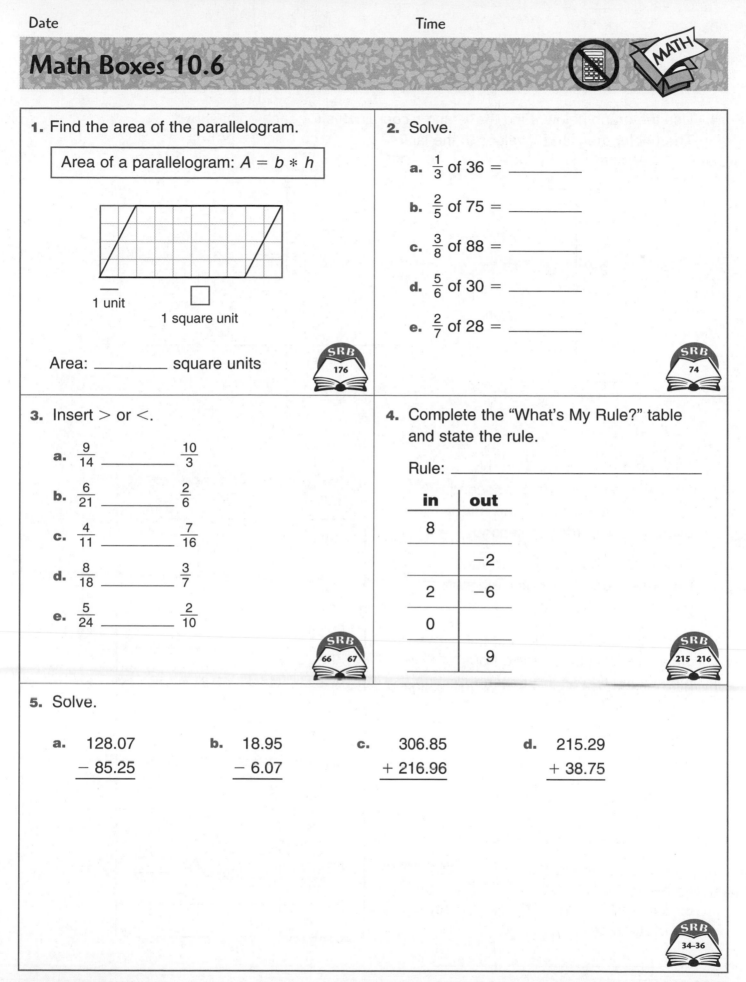

1. Find the area of the parallelogram.

Area of a parallelogram: $A = b * h$

1 unit

□ 1 square unit

Area: _____ square units

SRB 176

2. Solve.

a. $\frac{1}{3}$ of 36 = _____

b. $\frac{2}{5}$ of 75 = _____

c. $\frac{3}{8}$ of 88 = _____

d. $\frac{5}{6}$ of 30 = _____

e. $\frac{2}{7}$ of 28 = _____

SRB 74

3. Insert > or <.

a. $\frac{9}{14}$ _____ $\frac{10}{3}$

b. $\frac{6}{21}$ _____ $\frac{2}{6}$

c. $\frac{4}{11}$ _____ $\frac{7}{16}$

d. $\frac{8}{18}$ _____ $\frac{3}{7}$

e. $\frac{5}{24}$ _____ $\frac{2}{10}$

SRB 66 67

4. Complete the "What's My Rule?" table and state the rule.

Rule: _____

in	out
8	
	−2
2	−6
0	
	9

SRB 215 216

5. Solve.

a.
$$\begin{array}{r} 128.07 \\ -\ 85.25 \\ \hline \end{array}$$

b.
$$\begin{array}{r} 18.95 \\ -\ 6.07 \\ \hline \end{array}$$

c.
$$\begin{array}{r} 306.85 \\ +\ 216.96 \\ \hline \end{array}$$

d.
$$\begin{array}{r} 215.29 \\ +\ 38.75 \\ \hline \end{array}$$

SRB 34–36

Rules, Tables, and Graphs

1. Use the graph below. Find the *x*- and *y*-coordinates for each point shown. Then enter the *x* and *y* values in the table.

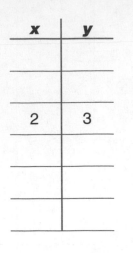

x	y
2	3

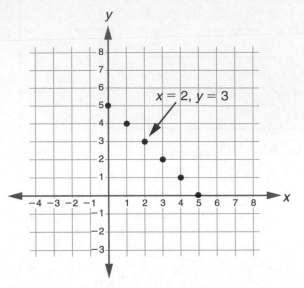

2. Eli is 10 years old and can run an average of 5 yards per second. His sister Sara is 7 and can run an average of 4 yards per second.

   Eli and Sara have a 60-yard race. Because Sara is younger, Eli gives her a 10-yard head start.

   Complete the table showing the distances Eli and Sara are from the starting line after 1 second, 2 seconds, 3 seconds, and so on.

   Use the table to answer the questions below.

   a. Who wins the race? _____

   b. What is the winning time?

   _____

   c. Who was in the lead during most of the race?

   _____

Time (sec)	Distance (yd) Eli	Distance (yd) Sara
start	0	10
1		
2		18
3	15	
4		
5		
6		
7		38
8		
9		
10		
11		
12		

Use with Lesson 10.6.

Rules, Tables, and Graphs (cont.)

3. Use the grid below to graph the results of the race between Eli and Sara.

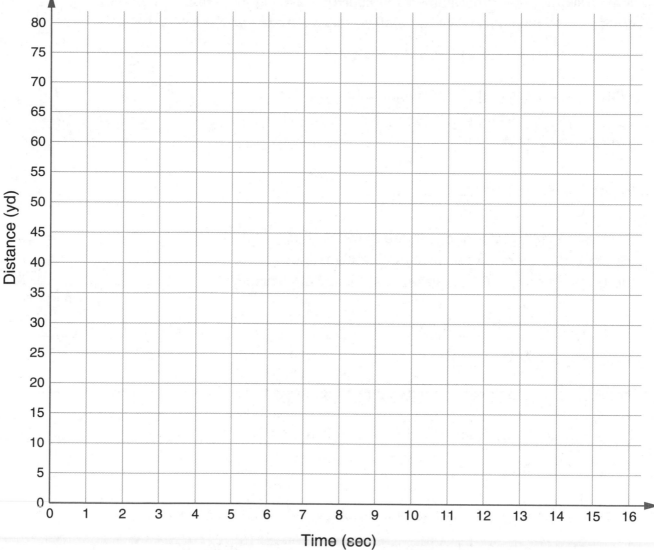

4. How many yards apart are Eli and Sara after 7 seconds?   _____

5. Suppose that Eli and Sara race for 75 yards instead of 60 yards.

   a. Who would you expect to win?   _____

   b. How long would the race last?   _____ seconds

   c. How far ahead would the winner be at the finish line?   _____ yards

Running and Walking Graph

Math Message

Rachel, William, and Tamara timed themselves traveling the same distance in different ways. Rachel ran, William walked, and Tamara walked toe-to-heel.

After they timed themselves, they drew a graph.

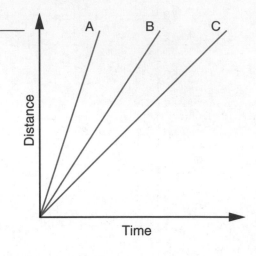

1. Which line on the graph at the right is for Rachel? _____

2. Which line is for William? _____

3. Which line is for Tamara? _____

4. Sam came along later and was the slowest of all. He walked heel-to-toe backwards. Draw a line on the graph to show at what speed you think Sam walked.

Review: Algebraic Expressions

Complete each statement with an algebraic expression.

5. Bill is 5 years older than Rick. If Rick is R years old,

   then Bill is _____ years old.

6. Rebecca's piano lesson is one half as long as Lisa's. If Lisa's piano lesson is

   L minutes long, then Rebecca's is _____ minutes long.

7. Jamie's dog weighs 3 pounds more than twice the weight of Eddy's dog.

   If Eddy's dog weighs E pounds, then Jamie's dog weighs _____ pounds.

Use with Lesson 10.7.

Reading Graphs

1. Tom and Alisha run a 200-yard race. Tom has a head start.

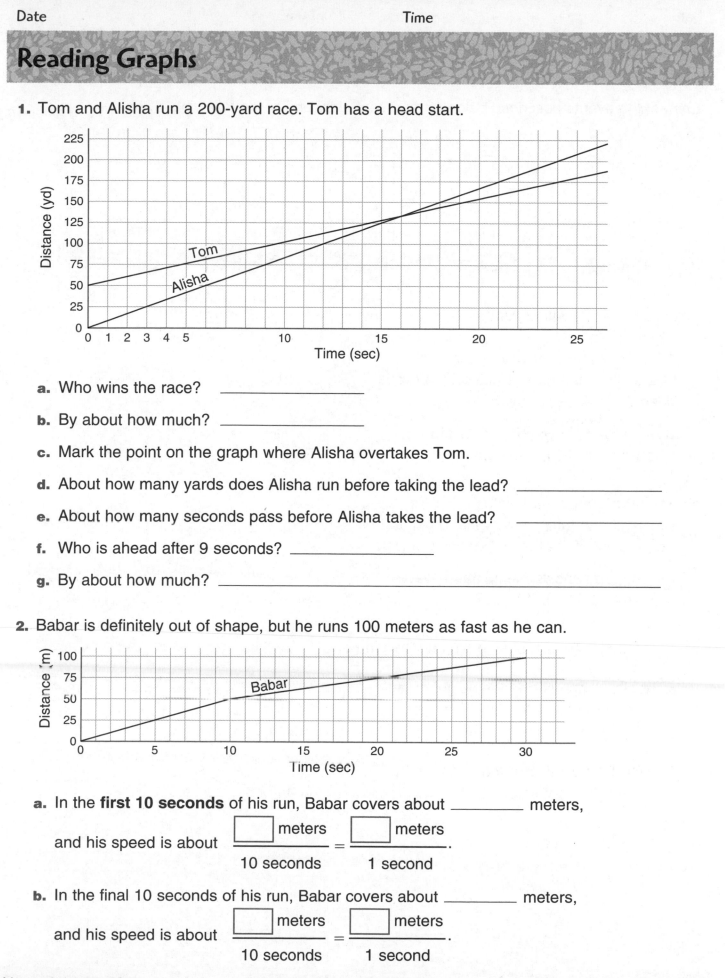

a. Who wins the race? _____

b. By about how much? _____

c. Mark the point on the graph where Alisha overtakes Tom.

d. About how many yards does Alisha run before taking the lead? _____

e. About how many seconds pass before Alisha takes the lead? _____

f. Who is ahead after 9 seconds? _____

g. By about how much? _____

2. Babar is definitely out of shape, but he runs 100 meters as fast as he can.

a. In the **first 10 seconds** of his run, Babar covers about _____ meters,

and his speed is about $\dfrac{\boxed{}\ \text{meters}}{10\ \text{seconds}} = \dfrac{\boxed{}\ \text{meters}}{1\ \text{second}}$.

b. In the final 10 seconds of his run, Babar covers about _____ meters,

and his speed is about $\dfrac{\boxed{}\ \text{meters}}{10\ \text{seconds}} = \dfrac{\boxed{}\ \text{meters}}{1\ \text{second}}$.

Mystery Graphs

Each of the events described below is represented by one of the following graphs:

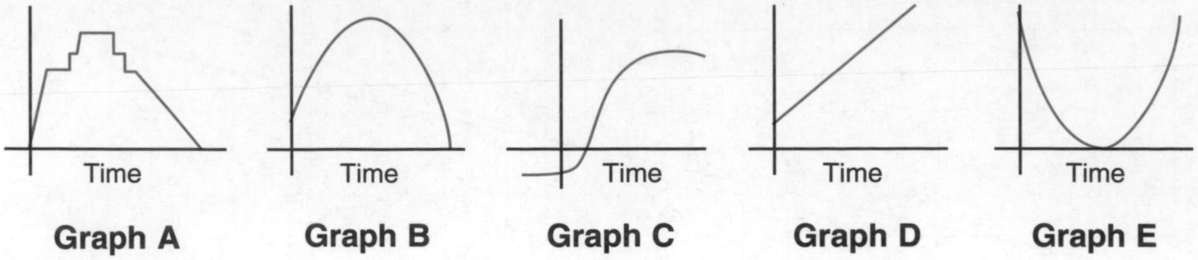

| Graph A | Graph B | Graph C | Graph D | Graph E |

Match each event with its graph.

1. A frozen dinner is removed from the freezer. It is heated in a microwave oven. Then it is placed on the table.

   Which graph shows the temperature
   of the dinner at different times?   Graph _____

2. Satya runs water into his bathtub. He steps into the tub, sits down, and bathes. He gets out of the tub and drains the water.

   Which graph shows the height of water
   in the tub at different times?   Graph _____

3. A baseball is thrown straight up into the air.

   a. Which graph shows the height of the
      ball—from the time it is thrown until
      the time it hits the ground?   Graph _____

   b. Which graph shows the speed of the
      ball at different times?   Graph _____

Use with Lesson 10.7.

Math Boxes 10.7

---

1. a. I am a polygon with exactly 4 angles, each of a different size. What shape am I?

_____

  b. Draw what I look like.

---

2. Add or subtract.

  a. $20 + (-10) =$ _____

  b. $-8 + (-17) =$ _____

  c. $-12 - (-12) =$ _____

  d. $-45 + 45 =$ _____

  e. $-31 - 14 =$ _____

---

3. List all of the factors for 144.

_____

_____

_____

---

4. Add or subtract.

  a. $5\frac{4}{5} - 3\frac{7}{4} =$ _____

  b. $3\frac{1}{8} + \frac{16}{6} =$ _____

  c. _____ $= 2\frac{4}{9} + 3\frac{7}{3}$

  d. _____ $= 1\frac{9}{10} - \frac{15}{8}$

---

5. Mr. Kim's art class asked 50 people each to name their favorite kind of movie. The results are shown in the table. Complete the table and then make a circle graph of the results.

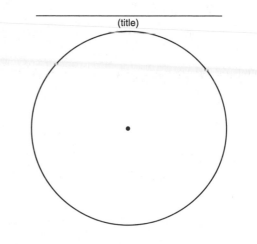

_____
(title)

Kind of Movie	Number of People	Percent of Total
Action	23	
Comedy	14	
Romance	2	
Thriller	7	
Mystery	4	
Total		

A Problem from the National Assessment

The following problem was in the mathematics section of a 1975 national standardized test.

A square has a perimeter of 12 inches.
What is the area of the square?

1. Your answer: _____ in.^{2}

The table below gives the national results for this problem.

	13-Year-Olds	**17-Year-Olds**	**Young Adults**
Correct answer	7%	28%	27%
144 sq inches	12%	19%	25%
48 sq inches	20%	10%	10%
24 sq inches	6%	4%	2%
12 sq inches	4%	3%	3%
6 sq inches	4%	2%	1%
3 sq inches	3%	2%	2%
Other incorrect answers	16%	13%	10%
No answer or "I don't know"	28%	19%	20%

Explain why many students might have given the following answers:

2. 144 square inches _____

_____

_____

3. 48 square inches _____

_____

_____

_____

Ratio of Circumference to Diameter

You are going to explore the relationship between the circumference and the diameter of a circle.

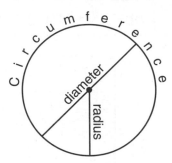

1. Using a metric tape measure, carefully measure the circumference and diameter of a variety of round objects. Measure to the nearest millimeter (one-tenth of a centimeter).

2. Record your data in the first three columns of the table below.

3. In the fourth column, write the ratio of the circumference to the diameter as a fraction.

4. In the fifth column, write the ratio as a decimal. Use your calculator to compute the decimal and round your answer to two decimal places.

Object	Circumference (C)	Diameter (d)	Ratio of Circumference to Diameter	
			Ratio as a Fraction ($\frac{C}{d}$)	Ratio as a Decimal (from calculator)
Coffee cup	252 mm	80 mm	$\frac{252}{80}$	3.15
	_____ mm	_____ mm		
	_____ mm	_____ mm		
	_____ mm	_____ mm		
	_____ mm	_____ mm		
	_____ mm	_____ mm		

5. What is the median of the circumference to diameter ratios in the last column?

   _____

6. The students in your class combined their results in a stem-and-leaf plot. Use that plot to find the class median value for the ratio $\frac{C}{d}$.

   _____

Math Boxes 10.8

1. Find the area of the triangle.

Area of a Triangle: $A = \frac{1}{2} * b * h$

$\overline{}$ 1 unit

☐ 1 square unit

Area: _____ square units

2. If a set has 48 objects, how many objects are there in

a. $\frac{3}{8}$ of the set? _____

b. $\frac{8}{3}$ of the set? _____

c. $\frac{5}{6}$ of the set? _____

d. $\frac{7}{12}$ of the set? _____

e. $\frac{17}{16}$ of the set? _____

3. Insert > or <.

a. $\frac{8}{9}$ _____ $\frac{8}{10}$

b. $\frac{3}{5}$ _____ $\frac{3}{7}$

c. $\frac{6}{7}$ _____ $\frac{5}{6}$

d. $\frac{7}{12}$ _____ $\frac{7}{14}$

e. $\frac{9}{11}$ _____ $\frac{14}{15}$

4. Complete the "What's My Rule?" table and state the rule.

Rule: _____

in	out
$\frac{1}{3}$	
	0
$\frac{5}{3}$	4
	2
−2	$\frac{1}{3}$

5. Solve.

a.   40.017
    + 269.000

b.   24.303
    + 5.700

c.   402.03
    − 24.70

d.   590.32
    − 465.75

Use with Lesson 10.8.

Measuring the Area of a Circle

Math Message

Use the circle at the right to solve Problems 1–4.

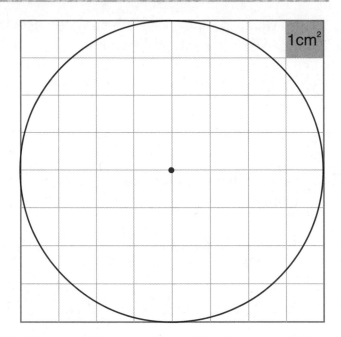

1cm²

1. The diameter of the circle is

   about _____ centimeters.

2. The radius of the circle is

   about _____ centimeters.

3. The circumference of the circle is about _____ centimeters.

4. Find the area of this circle by counting squares. About _____ cm²

Follow-Up

5. What is the median of all the area measurements in your class? _____ cm²

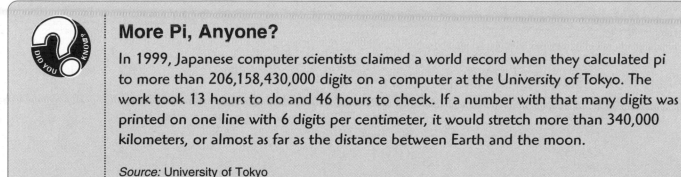

More Pi, Anyone?

In 1999, Japanese computer scientists claimed a world record when they calculated pi to more than 206,158,430,000 digits on a computer at the University of Tokyo. The work took 13 hours to do and 46 hours to check. If a number with that many digits was printed on one line with 6 digits per centimeter, it would stretch more than 340,000 kilometers, or almost as far as the distance between Earth and the moon.

Source: University of Tokyo

Areas of Circles

Work with a partner. Use the same objects, but make separate measurements so that you can check each other's work.

1. Trace several round objects onto the grid on *Math Masters,* page 2.

2. Count square centimeters to find the area of each circle.

3. Use a ruler to find the radius of each object. (*Remember:* The radius is half the diameter.) Record your data in the first three columns of the table below.

Object	Area (sq cm)	Radius (cm)	Ratio of Area to Radius Squared	
			as a Fraction $\dfrac{A}{r^2}$	as a Decimal

4. Find the ratio of the area to the square of the radius for each circle. Write the ratio as a fraction in the fourth column of the table. Then use a calculator to calculate the ratio as a decimal. Round the decimal to two decimal places and write it in the last column.

5. Find the median of the ratios in the last column. _____

Use with Lesson 10.9.

A Formula for the Area of a Circle

Your class just measured the area and radius of many circles and found that the ratio of the area to the square of the radius is about 3.

This was no coincidence: Mathematicians proved long ago that the ratio of the area of a circle to the square of its radius is always equal to π. This can be written as:

$$\frac{A}{r^2} = \pi$$

Usually this fact is written in a slightly different form, as a formula for the area of a circle.

> The formula for the area of a circle is
> $$A = \pi * r^2$$
> where A is the area of a circle and r is its radius.

1. What is the radius of the circle in the Math Message on journal page 373? _____

2. Use the formula above to calculate the area of that circle. _____

3. Is the area you found by counting square centimeters more or less than the

   area you found by using the formula? _____

   How much more or less? _____

4. Use the formula to find the areas of the circles you traced on *Math Masters,* page 2.

   _____   _____   _____

5. Which do you think is a more accurate way to find the area of a circle, by counting squares or by measuring the radius and using the formula? Explain.

   _____

   _____

   _____

   _____

Fraction and Percent Multiplication

Complete the following.

1. 30% of 50 is _____.

2. 25% of 36 is _____.

3. 5% of 150 is _____.

4. 75% of 12 is _____.

5. 80% of 60 is _____.

6. 50% of 130 is _____.

Find the whole.

7. 50% of _____ is 12.

8. $\frac{3}{4}$ of _____ is 21.

9. 90% of _____ is 180.

10. $\frac{5}{6}$ of _____ is 25.

11. 20% of _____ is 19.

12. $\frac{3}{8}$ of _____ is 24.

Multiply.

13. $\frac{1}{2} * \frac{3}{4} =$ _____

14. $2\frac{3}{4} * \frac{3}{5} =$ _____

15. $1\frac{1}{2} * 2\frac{1}{4} =$ _____

16. $\frac{3}{4} * 5 =$ _____

17. $7 * \frac{4}{5} =$ _____

18. $\frac{5}{6} * \frac{1}{5} =$ _____

Use with Lesson 10.9.

Math Boxes 10.9

1. I am a polygon with exactly 6 angles. What shape am I?

_____

Name an object that has my shape somewhere on it.

_____

Do all of my angles have to be the same size? _____ Explain. _____

_____

2. Add or subtract.

a. $24 + (-40) =$ _____

b. $-7 - (-23) =$ _____

c. $-43 + (-16) =$ _____

d. $37 - (-37) =$ _____

e. $-10 + 14 =$ _____

3. List all of the factors for 205.

_____

4. Add or subtract.

a. $3\frac{5}{8} - 1\frac{2}{5} =$ _____

b. $2\frac{6}{7} - \frac{9}{4} =$ _____

c. _____ $= \frac{37}{5} + 8\frac{3}{2}$

d. _____ $= \frac{12}{100} + \frac{25}{4}$

5. Ms. Hopheart's class asked 50 people to name their favorite kind of fruit. The results are shown in the table. Complete the table and then make a circle graph of the results.

Kind of Fruit	Number of People	Percent of Total
Apples	20	
Bananas	12	
Grapes	5	
Oranges	8	
Other	5	
Total	50	

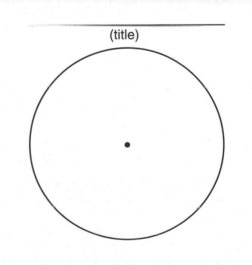

(title)

Time to Reflect

1. Explain to a new student what is special about the number π and what you have used it for.

_____

_____

_____

_____

_____

2. Look through your journal. Which lesson or lessons were your favorites in this unit? Explain.

_____

_____

_____

_____

_____

3. Janet wanted to know how long it would take her to drive 525 miles. She was traveling at about 65 miles per hour. Explain how she might solve this problem using a graph, a formula, or a table.

_____

_____

_____

_____

_____

 Use with Lesson 10.10.

Math Boxes 10.10

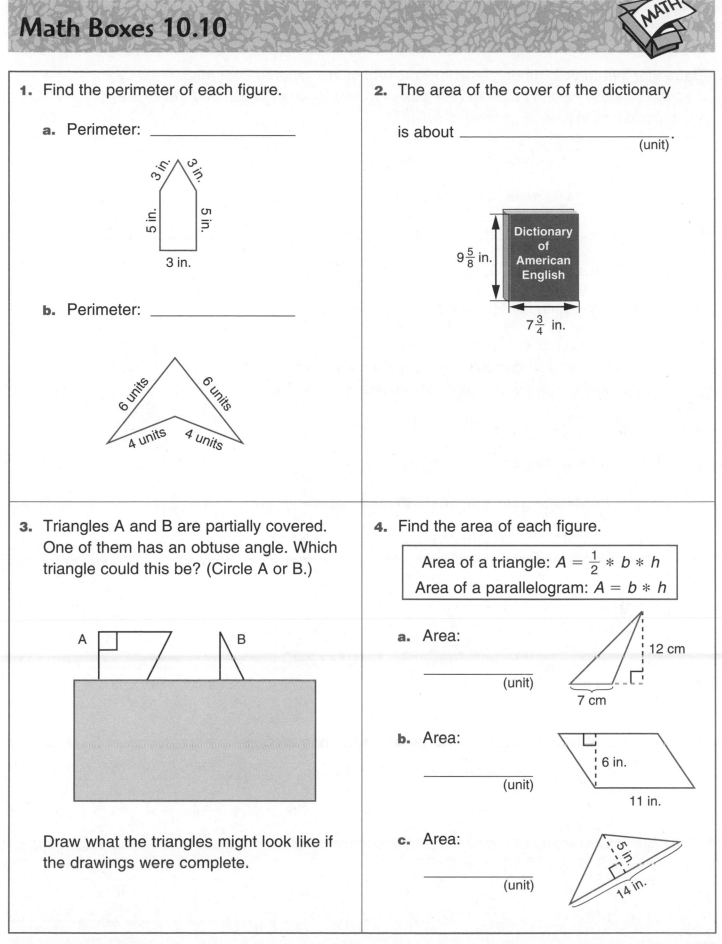

1. Find the perimeter of each figure.

  a. Perimeter: _____

3 in. 3 in.
5 in. 5 in.
3 in.

  b. Perimeter: _____

6 units   6 units
4 units   4 units

2. The area of the cover of the dictionary

is about _____ .
                              (unit)

$9\frac{5}{8}$ in.

Dictionary of American English

$7\frac{3}{4}$ in.

3. Triangles A and B are partially covered. One of them has an obtuse angle. Which triangle could this be? (Circle A or B.)

A   B

Draw what the triangles might look like if the drawings were complete.

4. Find the area of each figure.

Area of a triangle: $A = \frac{1}{2} * b * h$
Area of a parallelogram: $A = b * h$

  a. Area:

  _____
                  (unit)

12 cm
7 cm

  b. Area:

  _____
                  (unit)

6 in.
11 in.

  c. Area:

  _____
                  (unit)

5 in.
14 in.

Geometric Solids

Each member of your group should cut out one of the patterns from *Math Masters,*
pages 150–153. Fold the pattern and glue or tape it together. Then add this model to
your group's collection of geometric solids.

1. Examine your models of geometric solids.

    a. Which solids have all flat surfaces? _____

    _____

    b. Which have no flat surfaces? _____

    c. Which have both flat and curved surfaces? _____

    d. If you cut the label of a cylindrical can in a straight
    line perpendicular to the bottom, and then unrolled and
    flattened the label, what would be the shape of the label?

    _____

2. Examine your models of polyhedrons.

    a. Which polyhedrons have more faces than vertices? _____

    b. Which polyhedrons have the same number of faces and vertices? _____

    _____

    c. Which polyhedrons have fewer faces than vertices? _____

    _____

3. Examine your model of a cube.

    a. Does the cube have more edges than vertices, the same
    number of edges as vertices, or fewer edges than vertices? _____

    Is this true for all polyhedrons? _____ Explain. _____

    _____

    b. How many edges of the cube meet at each vertex? _____

    Is this true for all polyhedrons? _____ Explain. _____

    _____

Use with Lesson 11.1.

More Circumference and Area Problems

Circumference and Area of Circles Formulas

Circumference = π ∗ d   Area = π ∗ r^2

where *d* is the diameter and *r* is the radius of the circle.

Measure the diameter of the circle below to the nearest centimeter. Then use the
⬚π⬚ key on your calculator to solve these problems. If your calculator doesn't have
a ⬚π⬚ key, enter 3.14 each time you need π. Show answers to the nearest tenth.

1. The diameter of the circle is _____ cm.

2. The radius of the circle is _____ cm.

3. The circumference of the circle is _____ cm.

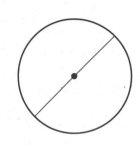

4. The area of the circle is _____ cm².

5. Explain the relationship between the diameter and the circumference. _____

_____

_____

6. Use your Geometry Template to draw
   a circle that has a diameter of 5 centimeters.

   a. Find the area of your circle.

      _____ cm²

   b. Find the circumference of your circle.

      _____ cm

7. Use your Geometry Template to draw
   a circle that has a radius of 1 inch.

   a. Find the area of your circle.

      _____ in.²

   b. Find the circumference of your circle.

      _____ in.

Math Boxes 11.1

---

1. Write the prime factorization for 200.

_____

SRB 12

2. If you draw one card from a regular deck of cards, what is the probability of drawing

  a. a 4? _____

  b. a face card? _____

  c. a heart? _____

  d. an even number? _____

SRB 122–123

---

3. Write an algebraic expression for each of the following statements.

  a. Maria is y years old. Sheila is 10 years older than Maria. How old is Sheila?

  _____ years old

  b. Franklin has c miniature cards. Rosie has 4 more cards than twice as many as Franklin has. How many cards does Rosie have? _____ cards

  c. Lucinda goes to sleep-away camp for D days each summer. Rhonda goes to camp for 1 day fewer than half of Lucinda's number of days. For how many days does Rhonda go to camp? _____ days

  d. Cheryl read B books this year. Ralph read 3 more than 5 times as many books as Cheryl. How many books did Ralph read? _____ books

SRB 202

---

4. Add or subtract.

  a. $\frac{3}{8} + \frac{5}{9} =$ _____

  b. $\frac{29}{4} + 1\frac{2}{5} =$ _____

  c. $\frac{18}{7} - 2\frac{1}{5} =$ _____

SRB 68–72

5. Solve.

  a. $3.26 + 504.1 =$ _____

  b. _____ $= 793.82 - 209.785$

  c. _____ $= 987.55 + 283.6$

SRB 34–36

Use with Lesson 11.1.

Math Boxes 11.2

1. Multiply. Do not use a calculator.

a. $\frac{3}{8} * \frac{5}{4} =$ _____

b. $\frac{2}{3} * \frac{6}{7} =$ _____

c. $\frac{1}{5} * \frac{8}{9} =$ _____

d. $1\frac{3}{4} * \frac{4}{5} =$ _____

SRB 76–78

2. Draw a circle with a radius of 1.5 centimeters.

$$\text{Area} = \pi * r^2$$

Find the area of the circle.

The area is about _____ cm^2.

SRB 154 178

3. Complete the table. Graph the data and connect the points with line segments.

Maryanne earns $12 per hour.

Rule:
earnings = 12 * hours

Hours	Earnings
2	
4	
	60
	84
9	

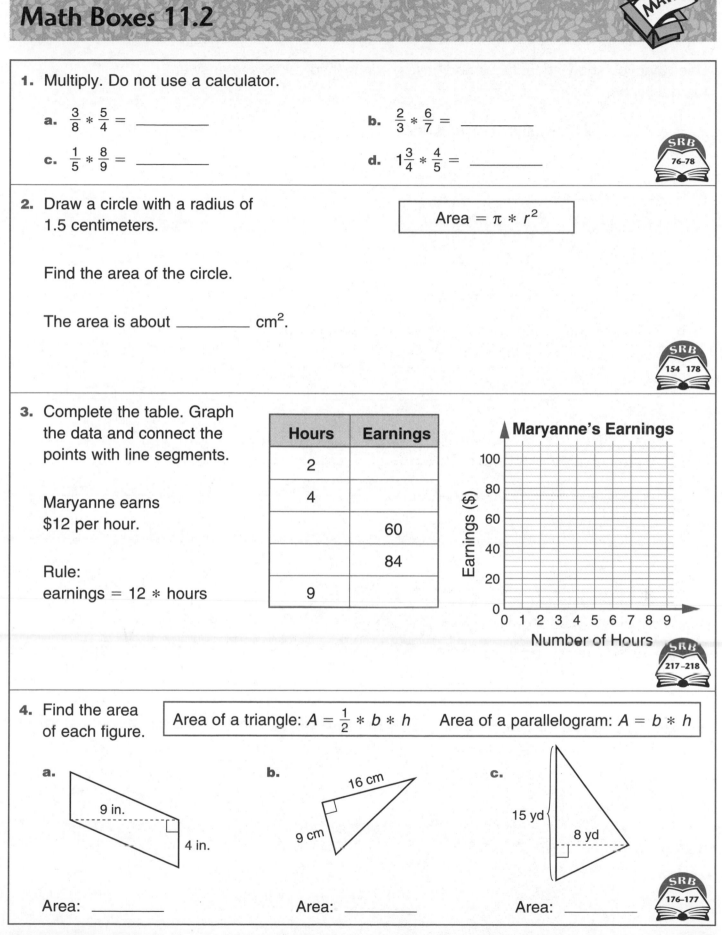

Maryanne's Earnings

Earnings ($)

Number of Hours

SRB 217–218

4. Find the area of each figure.

Area of a triangle: $A = \frac{1}{2} * b * h$ Area of a parallelogram: $A = b * h$

a.

9 in.

4 in.

b.

16 cm

9 cm

c.

15 yd

8 yd

Area: _____   Area: _____   Area: _____

SRB 176–177

Comparing Geometric Solids

Use what you know about faces and bases, edges, and vertices to answer the questions.

1. **a.** How are prisms and pyramids alike?

_____

_____

_____

_____

_____

b. What are some differences between prisms and pyramids?

_____

_____

_____

_____

_____

Prisms

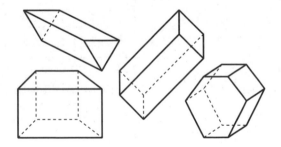

Pyramids

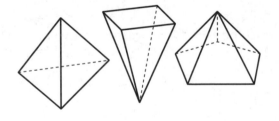

Cylinders

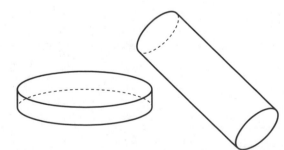

Cones

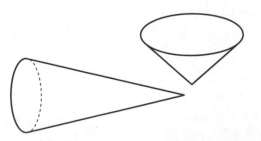

Use with Lesson 11.2.

Comparing Geometric Solids (cont.)

2. **a.** How are prisms and cylinders alike?

_____

_____

_____

_____

_____

  b. What are some differences between prisms and cylinders?

_____

_____

_____

_____

_____

3. **a.** How are pyramids and cones alike?

_____

_____

_____

_____

_____

  b. What are some differences between pyramids and cones?

_____

_____

_____

_____

_____

Volume of Cylinders

The base of a cylinder is circular. To find the area of the base of a cylinder, use the formula for finding the area of a circle.

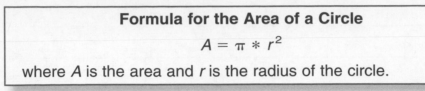

Formula for the Area of a Circle
$A = \pi * r^2$
where A is the area and r is the radius of the circle.

The formula for finding the volume of a cylinder is the same as the formula for finding the volume of a prism.

Formula for the Volume of a Cylinder
$V = B * h$
where V is the volume of the cylinder, B is the area of the base, and h is the height of the cylinder.

Use the two cans you have been given.

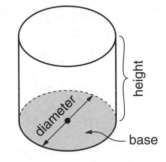

1. Measure the height of each can, inside the can. Measure the diameter of the base of each can. Record your measurements (to the nearest tenth of a centimeter) in the table below.

2. Calculate the radius of the base of each can. Then use the formula to find the volume. Record the results in the table.

3. Record the capacity of each can in the table, in milliliters.

	Height (cm)	Diameter of Base (cm)	Radius of Base (cm)	Volume (cm³)	Capacity (mL)
Can #1					
Can #2					

4. Measure the liquid capacity of each can. Fill the can with water. Then pour the water into a measuring cup. Keep track of the total amount of water you pour into the measuring cup.

Capacity of Can #1: _____ mL   Capacity of Can #2: _____ mL

Volume of Cylinders and Prisms

1. Find the volume of each cylinder.

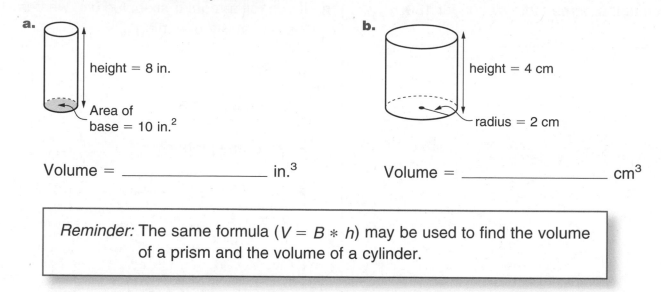

a.

height = 8 in.

Area of base = 10 in.²

Volume = _____ in.³

b.

height = 4 cm

radius = 2 cm

Volume = _____ cm³

> *Reminder:* The same formula ($V = B * h$) may be used to find the volume of a prism and the volume of a cylinder.

2. Find the volume of each wastebasket. Then determine which wastebasket has the largest capacity and which has the smallest.

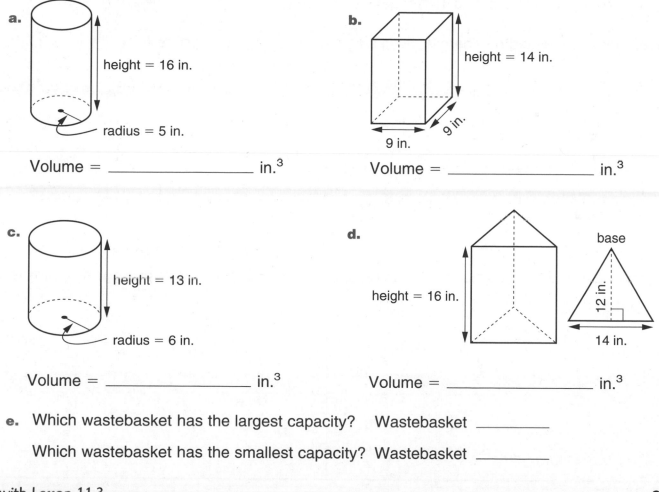

a.

height = 16 in.

radius = 5 in.

Volume = _____ in.³

b.

height = 14 in.

9 in.

9 in.

Volume = _____ in.³

c.

height = 13 in.

radius = 6 in.

Volume = _____ in.³

d.

height = 16 in.

base

12 in.

14 in.

Volume = _____ in.³

e. Which wastebasket has the largest capacity? Wastebasket _____

Which wastebasket has the smallest capacity? Wastebasket _____

Math Boxes 11.3

1. Write the prime factorization for 180.

_____

2. If you roll a regular six-sided die, what is the probability of getting

  a. a five? _____

  b. a prime number? _____

  c. an even number? _____

  d. a multiple of 3? _____

3. Theresa is y years old. Write an algebraic expression for the age of each person below.

  a. Nancy is four years older than Theresa. Nancy's age: _____ years

  b. Frank is twice as old as Theresa. Frank's age: _____ years

  c. José is $\frac{1}{3}$ as old as Theresa. José's age: _____ years

  d. Lucienne is 8 years younger than Theresa. Lucienne's age: _____ years

  e. If Theresa is 12, who is the oldest person above? _____

  How old is that person? _____

4. Add or subtract.

  a. $\begin{array}{r} 4\frac{2}{3} \\ -\ 3\frac{7}{8} \\ \hline \end{array}$
  b. $\begin{array}{r} 2\frac{14}{10} \\ +\ 1\frac{8}{9} \\ \hline \end{array}$
  c. $\begin{array}{r} 8\frac{20}{7} \\ -\ 6\frac{3}{9} \\ \hline \end{array}$

5. Solve.

  a. $\begin{array}{r} 52.6 \\ -\ 19.08 \\ \hline \end{array}$
  b. $\begin{array}{r} 703.93 \\ -\ 251.09 \\ \hline \end{array}$
  c. $\begin{array}{r} 826.3 \\ +\ 572.91 \\ \hline \end{array}$

Use with Lesson 11.3.

Volume of Pyramids and Cones

1. To calculate the volume of any **prism** or **cylinder,** you multiply the area of the base by the height. How would you calculate the volume of a **pyramid** or **cone**?

_____

_____

The Pyramid of Cheops is near Cairo, Egypt. It was built about 2600 B.C. It is a square pyramid. Each side of the square base is 756 feet long. Its height is 449.5 feet. The pyramid contains about 2,300,000 limestone blocks.

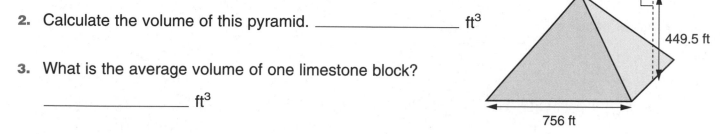

2. Calculate the volume of this pyramid. _____ ft³

3. What is the average volume of one limestone block?

_____ ft³

449.5 ft

756 ft

A movie theater sells popcorn in a box for $2.00. It also sells cones of popcorn for $1.50 each. The dimensions of the box and the cone are shown below.

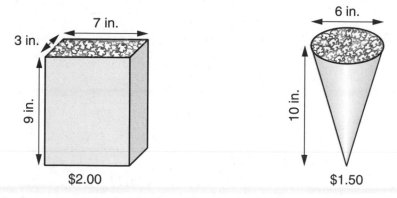

7 in.

3 in.

9 in.

$2.00

6 in.

10 in.

$1.50

4. Calculate the volume of the box. _____ in.³

5. Calculate the volume of the cone. _____ in.³

Challenge

6. Which is the better buy—the box or the cone of popcorn? Explain.

_____

_____

_____

Review of Area

Area Formulas		
Rectangle: $A = b * h$	Parallelogram: $A = b * h$	Triangle: $A = \frac{1}{2} * b * h$

where *A* is the area, *b* is the length of the base, and *h* is the height.

Find the areas of rectangles with the following dimensions. Do not forget the units.
You might want to make a sketch of the rectangles on a piece of scratch paper.

1. length of base = 8 in.   height = 15 in.   Area = _____   _____

2. length of base = 19 cm   height = 20 mm   Area = _____   _____

3. length of base = 18 in.   height = 3 ft   Area = _____   _____

Find the area of each of the polygons pictured below.

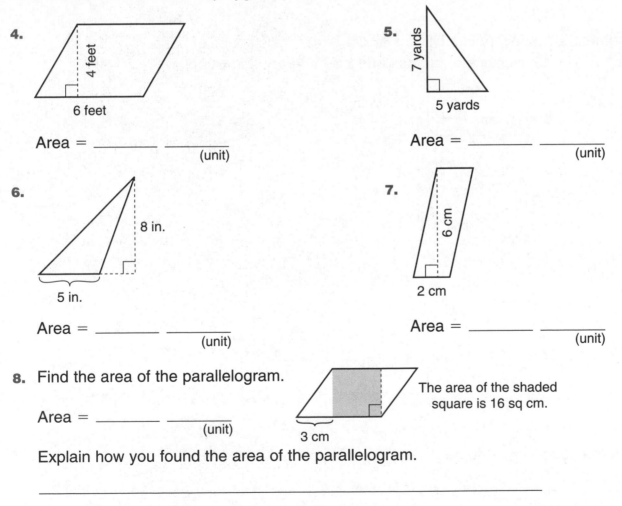

4.

4 feet

6 feet

Area = _____ _____
                        (unit)

5.

7 yards

5 yards

Area = _____ _____
                        (unit)

6.

8 in.

5 in.

Area = _____ _____
                        (unit)

7.

6 cm

2 cm

Area = _____ _____
                        (unit)

8. Find the area of the parallelogram.

The area of the shaded square is 16 sq cm.

3 cm

Area = _____ _____
                        (unit)

Explain how you found the area of the parallelogram.

_____

_____

_____

Use with Lesson 11.4.

Math Boxes 11.4

1. Multiply. Do not use a calculator.

a. $\frac{3}{8} * \frac{4}{7} =$ _____

b. $2\frac{2}{3} * 1\frac{3}{5} =$ _____

c. $1\frac{1}{8} * 2\frac{3}{4} =$ _____

d. $2\frac{1}{6} * 3\frac{1}{4} =$ _____

2. Allison's pizza has a radius of 8 inches.

Find the circumference of the pizza to

the nearest inch. _____

Find the area of the pizza to the nearest

square inch. _____

$$\text{Circumference} = \pi * d$$
$$\text{Area} = \pi * r^2$$

3. Complete the table.
Graph the data and
connect the points
with line segments.

Marissa runs at an
average speed of
6 miles per hour.

Rule: miles = 6 * hours

Hours	Miles
1	
2	
	30
6	
	48

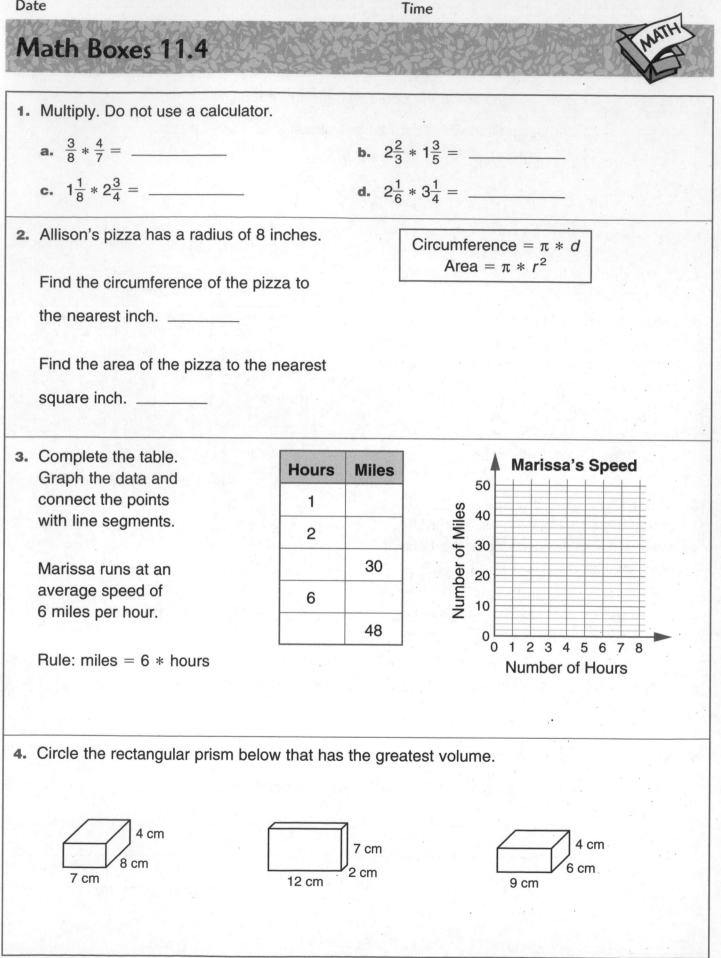

Marissa's Speed

4. Circle the rectangular prism below that has the greatest volume.

7 cm, 8 cm, 4 cm

12 cm, 2 cm, 7 cm

9 cm, 6 cm, 4 cm

How to Calibrate a Bottle

Materials
- ❏ 2-liter plastic soft-drink bottle with the top cut off
- ❏ can or jar filled with about 2 liters of water
- ❏ measuring cup
- ❏ ruler
- ❏ scissors
- ❏ paper
- ❏ tape

1. Fill the bottle with about 5 inches of water.

2. Cut a 1"-by-6" strip of paper. Tape the strip to the outside of the bottle with one end at the bottle top and the other end below the water level.

3. Mark the paper strip at the water level. Write "0 mL" next to the mark.

4. Pour 100 milliliters of water into a measuring cup. Pour the water into the bottle. Mark the paper strip at the new water level and write "100 mL."

5. Pour another 100 milliliters of water into the measuring cup. Pour it into the bottle and mark the new water level "200 mL."

6. Repeat, adding 100 milliliters at a time until the bottle is filled to within an inch of the top.

7. Pour out the water until the water level in the bottle falls to the 0-mL mark.

   How would you use your calibrated bottle to find the volume of a rock?

   _____

   _____

   _____

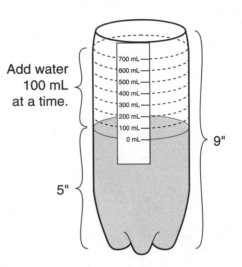

Add water
100 mL
at a time.

700 mL
600 mL
500 mL
400 mL
300 mL
200 mL
100 mL
0 mL

9"

5"

Use with Lesson 11.5.

Finding Volume by a Displacement Method

1. Check that the bottle is filled to the 0-mL level. Place several rocks in the bottle.

   Reminder: 1 mL = 1 cm^3

   a. What is the new level of the water in the bottle? _____ mL

   b. What is the volume of the rocks? _____ cm^3

   c. Does it matter whether the rocks are spread out or stacked? _____

2. Your fist has nearly the same volume as your heart. Here is a way to find the approximate volume of your heart. Check that the bottle is filled to the 0-mL level. Place a rubber band around your wrist, just below your wrist bone. Put your fist in the bottle until water reaches the rubber band.

   a. What is the new level of the water in the bottle? _____ mL

   b. What is the volume of your fist? This is the approximate volume of your heart. _____ cm^3

   c. Does it matter whether you make a fist or keep your hand open? _____

3. Find the volumes of several other objects in the same way. For example, find the volume of a baseball, a golf ball, an orange, an apple, or a full can of a soft drink. If the object floats in water, use a pencil to force it down. The object must be completely submerged before you read the water level.

Object	Volume of Water Object Displaces (mL)	Volume of Object (cm^3)

Fraction Review

Add or subtract.

1. $\frac{3}{8} + \frac{7}{8} =$ _____

2. $\frac{7}{12} - \frac{1}{8} =$ _____

3. $1\frac{1}{3} + \frac{5}{6} =$ _____

4. $\frac{5}{6} + \frac{1}{9} =$ _____

5. $\frac{2}{3} - \frac{3}{5} =$ _____

6. $3\frac{3}{7} + 1\frac{1}{2} =$ _____

7. $\frac{11}{12} - \frac{3}{4} =$ _____

8. $2\frac{5}{8} + 1\frac{1}{4} =$ _____

9. $3\frac{3}{7} - 1\frac{1}{2} =$ _____

10. You have 12 white tiles. If you add some black tiles so that the ratio
of white tiles to black tiles is 1 to 3, how many black tiles will you need? _____ black tiles

How many tiles will you have in all? _____ tiles

11. You have 15 white tiles. If you add some black tiles so that
3 out of 4 tiles are white, how many black tiles will you need? _____ black tiles

How many tiles will you have in all? _____ tiles

12. You have a total of 24 tiles. Five out of 8 tiles are black.

How many black tiles do you have? _____ black tiles

Use with Lesson 11.5.

Math Boxes 11.5

1. Solve.

a. $\frac{1}{2}$ of 12 = _____

b. $\frac{2}{3}$ of 18 = _____

c. $\frac{3}{8}$ of 24 = _____

d. $\frac{6}{9}$ of 30 = _____

e. $\frac{1}{2}$ of $\frac{1}{2}$ = _____

SRB 74–75

2. Draw 12 shaded tiles. Then draw some unshaded tiles so that 3 out of 5 tiles are shaded.

SRB 102

3. Solve the pan-balance problems below.

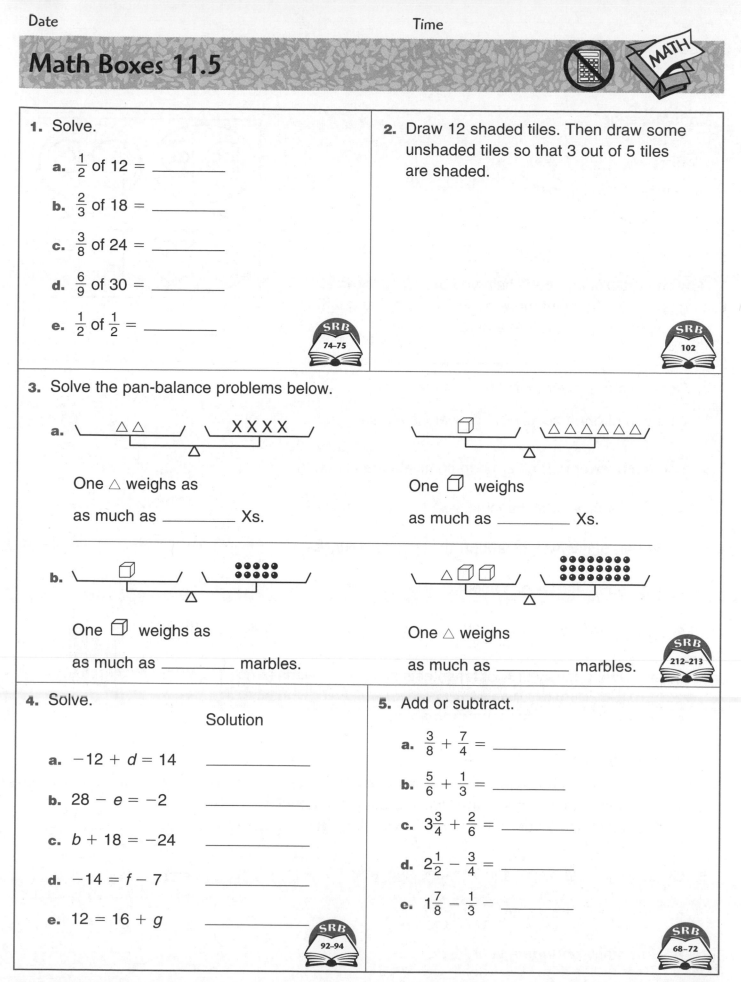

a.   △△   X X X X

One △ weighs as

as much as _____ Xs.

One ▢ weighs

as much as _____ Xs.

b.   ▢

One ▢ weighs as

as much as _____ marbles.

One △ weighs

as much as _____ marbles.

SRB 212–213

4. Solve.

                        Solution

a. $-12 + d = 14$ _____

b. $28 - e = -2$ _____

c. $b + 18 = -24$ _____

d. $-14 = f - 7$ _____

e. $12 = 16 + g$ _____

SRB 92–94

5. Add or subtract.

a. $\frac{3}{8} + \frac{7}{4} =$ _____

b. $\frac{5}{6} + \frac{1}{3} =$ _____

c. $3\frac{3}{4} + \frac{2}{6} =$ _____

d. $2\frac{1}{2} - \frac{3}{4} =$ _____

e. $1\frac{7}{8} - \frac{1}{3} -$ _____

SRB 68–72

Capacity and Weight

Math Message

1 pint = _____ cups

1 quart = _____ pints

1 half-gallon = _____ quarts

1 gallon = _____ quarts

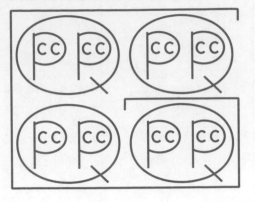

How can the picture above help you remember how many cups are in a pint, how many pints are in a quart, and how many quarts are in a gallon?

---

1. Round your answer to the nearest ounce.

   One cup of dry (uncooked) rice weighs about _____ ounces.

2. Use the answer in Problem 1 to complete the following.

   a. 1 pint of rice weighs about _____ ounces.

   b. 1 quart of rice weighs about _____ ounces.

   c. 1 gallon of rice weighs about _____ ounces.

   d. 1 gallon of rice weighs about _____ pounds. (1 pound = 16 ounces)

3. On average, a family of 4 in Japan eats about 40 pounds of rice a month.

   a. That's about how many **pounds** a year? _____

   b. How many **gallons** a year? _____

4. On average, a family of 4 in the United States eats about 88 pounds of rice a year. That's about how many gallons a year? _____

5. On average, a family of 4 in Thailand eats about 3 gallons of rice a week.

   a. That's about how many **gallons** a year? _____

   b. How many **pounds** a year? _____

Use with Lesson 11.6.

Capacity and Weight (cont.)

6. Find the capacity of the copy-paper carton shown at the right.

_____ in.³

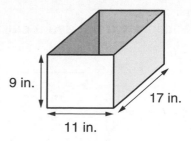

9 in.

17 in.

11 in.

7. The container at the right is a $\frac{1}{2}$-gallon juice container with the top cut off so that $\frac{1}{2}$ gallon of juice exactly fills it.

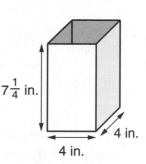

a. Find the volume of the $\frac{1}{2}$-gallon container.   _____ in.³

b. What is the volume of a 1-gallon container?   _____ in.³

$7\frac{1}{4}$ in.

4 in.

4 in.

8. On average, a family of 4 in Thailand eats about 156 gallons of rice a year. About how many copy-paper cartons will you need to hold this amount of rice? (*Hint:* First calculate how many gallons of rice will fill 1 copy-paper carton.)

a. What is the capacity of 1 copy-paper carton?

About _____ gallons

b. How many copy-paper cartons will you need to hold 156 gallons of rice?

About _____ cartons

Challenge

9. Estimate about how many pounds a copy-paper carton full of rice weighs. Describe what you did to find your estimate.

_____

_____

_____

_____

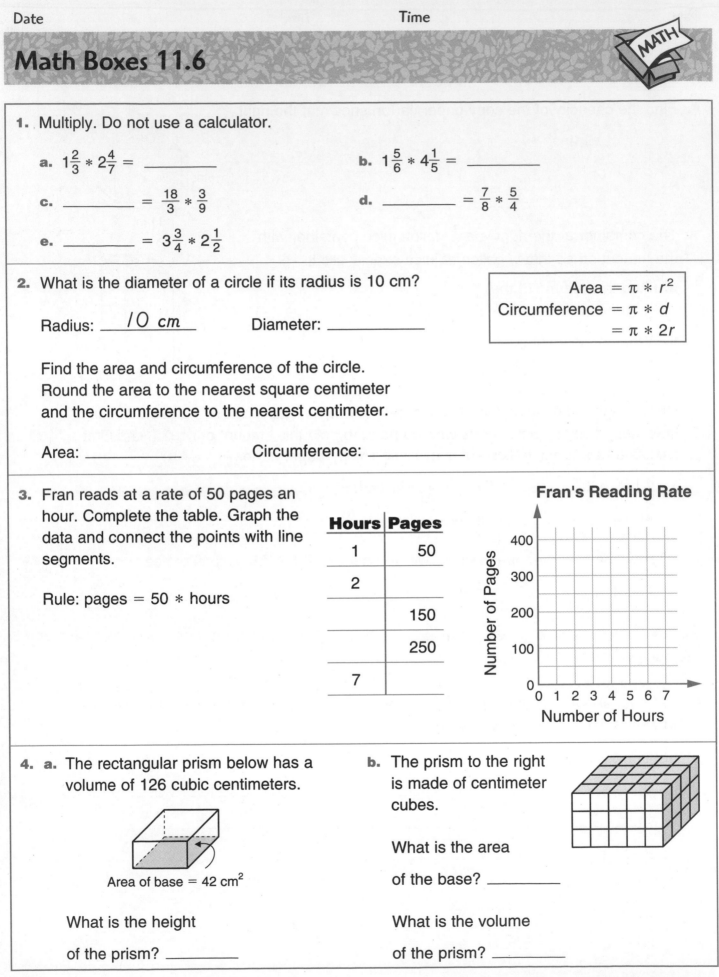

1. Multiply. Do not use a calculator.

   a. $1\frac{2}{3} * 2\frac{4}{7} =$ _____

   b. $1\frac{5}{6} * 4\frac{1}{5} =$ _____

   c. _____ $= \frac{18}{3} * \frac{3}{9}$

   d. _____ $= \frac{7}{8} * \frac{5}{4}$

   e. _____ $= 3\frac{3}{4} * 2\frac{1}{2}$

2. What is the diameter of a circle if its radius is 10 cm?

Area $= \pi * r^2$
Circumference $= \pi * d$
$= \pi * 2r$

   Radius: __*10 cm*__   Diameter: _____

   Find the area and circumference of the circle.
   Round the area to the nearest square centimeter
   and the circumference to the nearest centimeter.

   Area: _____   Circumference: _____

3. Fran reads at a rate of 50 pages an hour. Complete the table. Graph the data and connect the points with line segments.

   Rule: pages = 50 * hours

Hours	Pages
1	50
2	
	150
	250
7	

   Fran's Reading Rate

4. a. The rectangular prism below has a volume of 126 cubic centimeters.

   Area of base = 42 cm^2

   What is the height

   of the prism? _____

   b. The prism to the right is made of centimeter cubes.

   What is the area

   of the base? _____

   What is the volume

   of the prism? _____

Use with Lesson 11.6.

Math Boxes 11.7

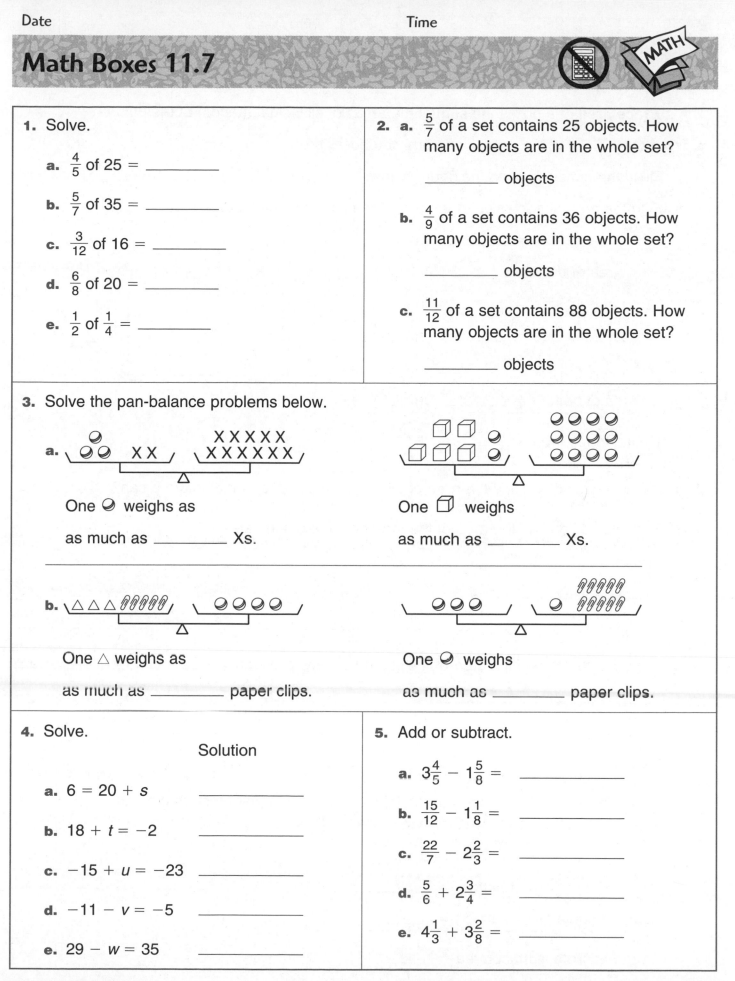

1. Solve.

  a. $\frac{4}{5}$ of 25 = _____

  b. $\frac{5}{7}$ of 35 = _____

  c. $\frac{3}{12}$ of 16 = _____

  d. $\frac{6}{8}$ of 20 = _____

  e. $\frac{1}{2}$ of $\frac{1}{4}$ = _____

2. a. $\frac{5}{7}$ of a set contains 25 objects. How many objects are in the whole set?

  _____ objects

  b. $\frac{4}{9}$ of a set contains 36 objects. How many objects are in the whole set?

  _____ objects

  c. $\frac{11}{12}$ of a set contains 88 objects. How many objects are in the whole set?

  _____ objects

3. Solve the pan-balance problems below.

  a.

  One ◯ weighs as

  as much as _____ Xs.

  One ▱ weighs

  as much as _____ Xs.

  b.

  One △ weighs as

  as much as _____ paper clips.

  One ◯ weighs

  as much as _____ paper clips.

4. Solve.

  Solution

  a. $6 = 20 + s$ _____

  b. $18 + t = -2$ _____

  c. $-15 + u = -23$ _____

  d. $-11 - v = -5$ _____

  e. $29 - w = 35$ _____

5. Add or subtract.

  a. $3\frac{4}{5} - 1\frac{5}{8} =$ _____

  b. $\frac{15}{12} - 1\frac{1}{8} =$ _____

  c. $\frac{22}{7} - 2\frac{2}{3} =$ _____

  d. $\frac{5}{6} + 2\frac{3}{4} =$ _____

  e. $4\frac{1}{3} + 3\frac{2}{8} =$ _____

Surface Area

The **surface area** of a box is the sum of the areas of all 6 sides (faces) of the box.

1. Your class will find the dimensions of a cardboard box.

a. Fill in the dimensions on the figure below.

b. Find the area of each side of the box. Then find the total surface area.

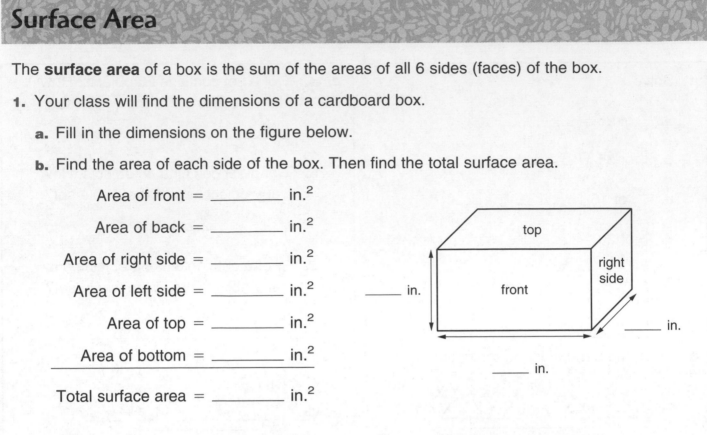

Area of front = _____ in.^{2}

Area of back = _____ in.^{2}

Area of right side = _____ in.^{2}

Area of left side = _____ in.^{2}

Area of top = _____ in.^{2}

Area of bottom = _____ in.^{2}

Total surface area = _____ in.^{2}

2. *Think:* How would you find the **area** of all the metal used to manufacture a can?

a. How would you find the area of the top or bottom of the can?

_____

_____

_____

b. How would you find the area of the curved surface between the top and bottom of the can?

_____

_____

_____

c. Choose a can. Find the total area of the metal used to manufacture the can. Remember to include a unit for each area.

Area of top = _____

Area of bottom = _____

Area of curved side surface = _____

Total surface area = _____

Use with Lesson 11.7.

Surface Area (cont.)

Formula for the Area of a Triangle

$$A = \frac{1}{2} * b * h$$

where A is the area of the triangle, b is the length of its base, and h is its height.

---

3. Use your model of a triangular prism.

a. Find the dimensions of the triangular and rectangular faces. Then find the areas of these faces. Measure lengths to the nearest $\frac{1}{4}$ inch.

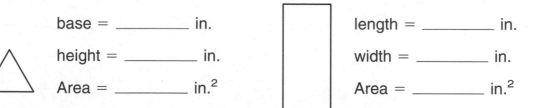

base = _____ in.   length = _____ in.

height = _____ in.   width = _____ in.

Area = _____ in.²   Area = _____ in.²

b. Add the areas of the faces to find the total surface area.

Area of 2 triangular bases = _____ in.²

Area of 3 rectangular sides = _____ in.²

Total surface area = _____ in.²

4. Use your model of a square pyramid.

a. Find the dimensions of the square and triangular faces. Then find the areas of these faces. Measure lengths to the nearest tenth of a centimeter.

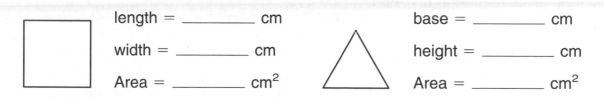

length = _____ cm   base = _____ cm

width = _____ cm   height = _____ cm

Area = _____ cm²   Area = _____ cm²

b. Add the areas of the faces to find the total surface area.

Area of square base = _____ cm²

Area of 4 triangular sides = _____ cm²

Total surface area = _____ cm²

Time to Reflect

1. Explain what is meant by the volume of a three-dimensional object. Pretend that you are trying to explain it to a new student.

_____

_____

_____

_____

2. Describe at least two situations in which you would find the capacity of something.

_____

_____

_____

_____

3. Describe at least two situations in which you would find the surface area of something.

_____

_____

_____

_____

4. Look back through your journal. List at least one concept that you studied in this unit that you think you will find useful.

_____

_____

_____

_____

5. What was your favorite lesson in this unit? Explain.

_____

_____

_____

Math Boxes 11.8

1. Write the prime factorization for 175.

_____

2. Solve.

a. $\frac{3}{8}$ of 40 = _____

b. $\frac{2}{3}$ of 120 = _____

c. $\frac{4}{5}$ of 60 = _____

d. $\frac{7}{9}$ of 54 = _____

e. $\frac{5}{6}$ of 36 = _____

3. Multiply.

a. $\frac{7}{8} * \frac{8}{9}$ = _____

b. _____ $= 1\frac{1}{3} * 2\frac{1}{5}$

c. _____ $= 4\frac{1}{6} * 3\frac{1}{3}$

d. _____ $= \frac{25}{6} * \frac{8}{9}$

e. _____ $= 5 * 2\frac{5}{7}$

4. Add.

a. $\frac{3}{8} + \frac{2}{5}$ = _____

b. $\frac{5}{6} + \frac{3}{4}$ = _____

c. $1\frac{4}{7} + \frac{2}{3}$ = _____

d. $5\frac{5}{9} + 2\frac{1}{7}$ = _____

e. $4\frac{1}{5} + 1\frac{7}{8}$ = _____

5. Color the spinner so that there is a 25% chance of landing on red and a $\frac{1}{3}$ chance of landing on black. Leave the rest of the spinner white.

What is the probability
of landing on white? _____

If you spin the spinner 300 times, about how many times would you expect the spinner to land on black?

_____

Math Boxes 12.1

1. Solve.

a. If 15 marbles are $\frac{3}{5}$ of the marbles in a
bag, how many marbles are in the bag? _____ marbles

b. If 14 pennies are 7% of a pile of
pennies, how many pennies are in the pile? _____ pennies

c. 75 students are absent today. This is 10% of the students enrolled
at the school. How many students are enrolled at the school? _____ students

d. Jane paid $90 for a new radio. It was on sale for $\frac{3}{4}$ of
the regular price. What is the regular price of the radio? _____

SRB
52–53
75

2. Name the number for each point marked on the number line.

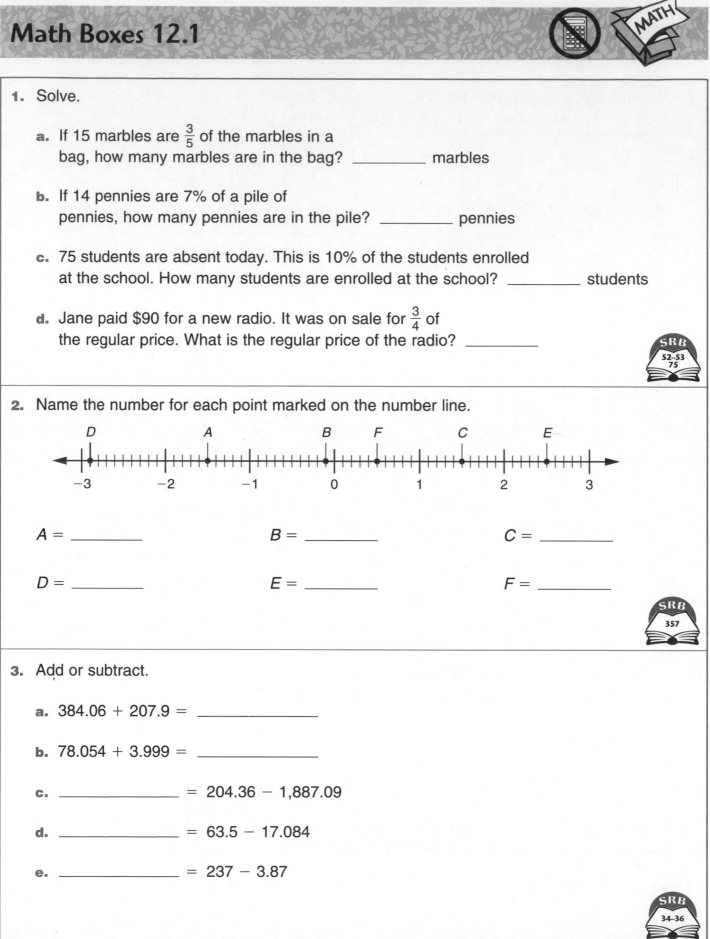

$A =$ _____ $B =$ _____ $C =$ _____

$D =$ _____ $E =$ _____ $F =$ _____

SRB
357

3. Add or subtract.

a. $384.06 + 207.9 =$ _____

b. $78.054 + 3.999 =$ _____

c. _____ $= 204.36 - 1{,}887.09$

d. _____ $= 63.5 - 17.084$

e. _____ $= 237 - 3.87$

SRB
34–36

Factors

Math Message

1. Write all the pairs of factors whose product is 48. One pair has been done for you.

   48 = $6 * 8$; _____

2. One way to write 36 as a product of factors is 2 * 18. Another way is 2 * 2 * 9. Write 36 as the product of the longest possible string of factors. Do not include 1 as a factor.

   _____

   _____

Factor Trees

One way to find all the prime factors of a number is to make a **factor tree.** First, write the number. Then, underneath, write any two factors whose product is that number. Then write factors of each of these factors. Continue until all the factors are prime numbers. Below are three factor trees for 36.

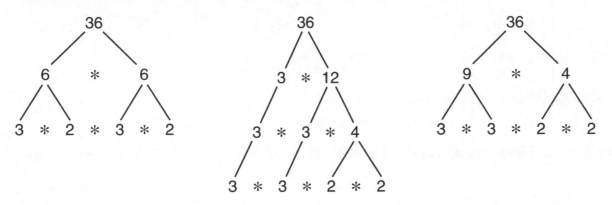

It does not matter which two factors you begin with. You always end with the same prime factors; for 36, they are 2, 2, 3, and 3. The **prime factorization** of 36 is 2 * 2 * 3 * 3.

3. Make a factor tree for each number. Then write the prime factorization.

   a.   24   b.   50

   24 = _____   50 = _____

Factor Trees and Greatest Common Factors

The **greatest common factor** of two whole numbers is the largest number that is a factor of both numbers.

Example 1 Find the greatest common factor of 24 and 60.

Step 1 List all the factors of 24: 1, 2, 3, 4, 6, 8, 12, and 24.

Step 2 List all the factors of 60: 1, 2, 3, 4, 5, 6, 10, 12, 15, 20, 30, and 60.

Step 3 1, 2, 3, 4, 6, and 12 are on both lists. They are **common factors.**
12 is the largest number. It is the greatest common factor of 24 and 60.

Another way to find the greatest common factor of two numbers is to use prime factorization.

Example 2 Find the greatest common factor of 24 and 60.

Step 1 Write the prime factorization of each number.

$$24 = 2 * 2 * 2 * 3 \qquad\qquad 60 = 2 * 2 * 3 * 5$$

Step 2 Circle pairs of common factors.

$$24 = 2 * 2 * 2 * 3$$
$$60 = 2 * 2 * 3 * 5$$

Step 3 Multiply *one* factor *in each pair* of circled factors.

The greatest common factor of 24 and 60 is 2 * 2 * 3, or 12.

1. Make a factor tree for each number below.

   a.   10   b.   75   c.   90

Factor Trees and Greatest Common Factors (cont.)

2. **a.** Which prime factors do 10 and 75 have in common? _____

     b. What is the greatest common factor of 10 and 75? _____

3. **a.** Which prime factors do 75 and 90 have in common? _____

     b. What is the greatest common factor of 75 and 90? _____

4. **a.** Which prime factors do 10 and 90 have in common? _____

     b. What is the greatest common factor of 10 and 90? _____

5. Use the factor trees in Problem 1 to help you write each fraction below in simplest form. Divide the numerator and denominator by their greatest common factor.

a. $\dfrac{10}{75} =$ _____

b. $\dfrac{75}{90} =$ _____

c. $\dfrac{10}{90} =$ _____

6. What is the greatest common factor of 20 and 25?
(*Hint:* Use factor trees to help you.) _____

Write the fraction $\dfrac{20}{25}$ in simplest form. $\qquad \dfrac{20}{25} =$ _____

Challenge

7. What is the greatest common factor of 1,260 and 1,350? _____

(*Hint:* $1{,}260 = 2 * 2 * 3 * 3 * 5 * 7$ and $1{,}350 = 2 * 3 * 3 * 3 * 5 * 5$.)

Factor Trees and Least Common Multiples

The **least common multiple** of two numbers is the smallest number that is a multiple of both numbers.

Example Find the least common multiple of 8 and 12.

Step 1 List the multiples of 8: 8, 16, 24, 32, 40, 48, 56, and so on.

Step 2 List the multiples of 12: 12, 24, 36, 48, 60, and so on.

Step 3 24 and 48 are in both lists. They are common multiples.
24 is the smallest number. It is the least common multiple for 8 and 12.
24 is also the smallest number that can be divided by both 8 and 12.

Another way to find the least common multiple for two numbers is to use prime factorization.

Example Find the least common multiple of 8 and 12.

Step 1 Write the prime factorization of each number:

$8 = 2 * 2 * 2$ $12 = 2 * 2 * 3$

Step 2 Circle pairs of common factors. Then cross out one factor in each pair as shown below.

$8 = 2 * 2 * 2$
$12 = 2 * 2 * 3$

Step 3 Multiply the factors that are not crossed out. The least common multiple of 8 and 12 is $2 * 2 * 2 * 3$, or 24.

1. Make factor trees and write the prime factorizations for each number.

   a.   15   **b.**   9   **c.**   30

   15 = _____   9 = _____   30 = _____

2. What is the least common multiple of …

   a. 9 and 15? _____   **b.** 15 and 30? _____   **c.** 9 and 30? _____

Use with Lesson 12.1.

Rate Number Stories

1. Mica reads about 44 pages in an hour.
About how many pages will she read in $2\frac{3}{4}$ hour? _____ pages

Explain how you found your answer. _____

_____

If Mica starts reading a 230-page book at 3:30 P.M., and she reads straight
through the book (without stopping), about what time will Mica finish the book? _____

Explain how you found your answer. _____

_____

_____

_____

_____

_____

_____

2. Tyree and Jake built a tower of centimeter cubes. The bottom floor of the tower is
rectangular. It is 5 cubes wide and 10 cubes long. The completed tower is the
shape of a rectangular prism. They began building at 2 P.M. They built for about
1 hour. They used approximately 200 cubes every 10 minutes.

How tall was the final tower? _____
                                    (unit)

Explain how you found your answer. _____

_____

_____

_____

_____

_____

_____

_____

Probability

When a fair 6-sided die is rolled, each number from 1 to 6 has an equal chance of coming up. The numbers 1, 2, 3, 4, 5, and 6 are **equally likely.**

The spinner below is divided into 10 equal sections. There is an equal chance of spinning each number from 1 through 10. The numbers 1, 2, 3, ..., 9, 10 are **equally likely.** This does not mean that if you spin 10 times, each number from 1 to 10 will come up exactly once. A 2 might come up four times, and a 10 might not come up at all. But if you spin many times (say 1,000 times), each number is likely to come up about $\frac{1}{10}$ of the time. The **probability** of landing on 1 is $\frac{1}{10}$. The probability of landing on 2 is also $\frac{1}{10}$, and so on.

Example What is the probability that the spinner at the right will land on an even number?

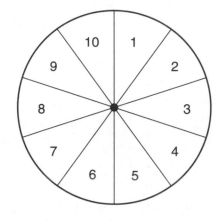

The spinner will land on an even number if it lands on 2, 4, 6, 8, or 10. Each of these even numbers is likely to come up $\frac{1}{10}$ of the time. The total probability that one of these even numbers will come up is found by adding:

$$\frac{1}{10} + \frac{1}{10} + \frac{1}{10} + \frac{1}{10} + \frac{1}{10} = \frac{5}{10}$$

Lands on   2   4   6   8   10

The probability of landing on an even number is $\frac{5}{10}$.
Find the probability of each of the following for this spinner.

1. The spinner lands on an odd number. _____

2. The spinner lands on a number less than 7. _____

3. The spinner lands on a multiple of 3. _____

4. The spinner lands on a number that is a factor of 12. _____

5. The spinner lands on the greatest common factor of 4 and 6. _____

6. The spinner lands on a prime number. _____

7. The spinner lands on a number that is NOT a prime number. _____

Use with Lesson 12.2.

The Multiplication Counting Principle and Tree Diagrams

Multiplication Counting Principle

Suppose you can make a first choice in m ways, and a second choice in n ways. Then there are $m * n$ ways of making the first choice followed by the second choice. Three or more choices can be counted in the same way, by multiplying.

A school cafeteria offers these choices for lunch:

Main Course: chili or hamburger

Drink: milk or juice

Dessert: apple or cake

1. a. How many different ways can a student choose one main course, one drink, and one dessert? Use the Multiplication Counting Principle.

_____   *   _____   *   _____

    (ways to choose   (ways to choose   (ways to choose

     a main course)   a drink)   a dessert)

b. Number of different ways to select foods for lunch: _____

2. Draw a **tree diagram** to show all possible ways to select foods for lunch.

Main Course: _____   _____

Drink: _____   _____   _____   _____

Dessert: _____ _____ _____ _____ _____ _____ _____ _____

3. a. Do you think that all of the ways to select foods for lunch are equally likely? _____

b. Explain your answer. _____

_____

_____

_____

Tree Diagrams and Probability

Sam has 3 clean shirts (red, blue, and yellow) and 2 clean pairs of pants (tan and black). He grabs a shirt and a pair of pants without looking.

1. Complete the tree diagram to show all possible ways that Sam can grab a shirt and a pair of pants.

Shirts: _____   _____   _____

Pants: _____ _____   _____ _____   _____ _____

2. List all possible combinations of shirts and pants. One has been done for you.

red and black, _____

_____

3. How many different combinations of shirts and pants are there? _____ combinations

4. Are all the shirt-pants combinations equally likely? _____

5. What is the probability that Sam will grab the following?

   a. the blue shirt _____

   b. the blue shirt and the black pants _____

   c. the tan pants _____

   d. a shirt that is NOT yellow _____

   e. the tan pants and a shirt that is NOT yellow _____

                                                   Use with Lesson 12.2.

Tree Diagrams and Probability (cont.)

Mr. Jackson travels to and from work by train. Trains to work leave at 6:00, 7:00, 8:00, and 9:00 A.M. Trains from work leave at 3:00, 4:00, and 5:00 P.M.

Mr. Jackson is equally likely to select any 1 of the 4 morning trains to go to work.

He is equally likely to select any of the 3 afternoon trains to go home from work.

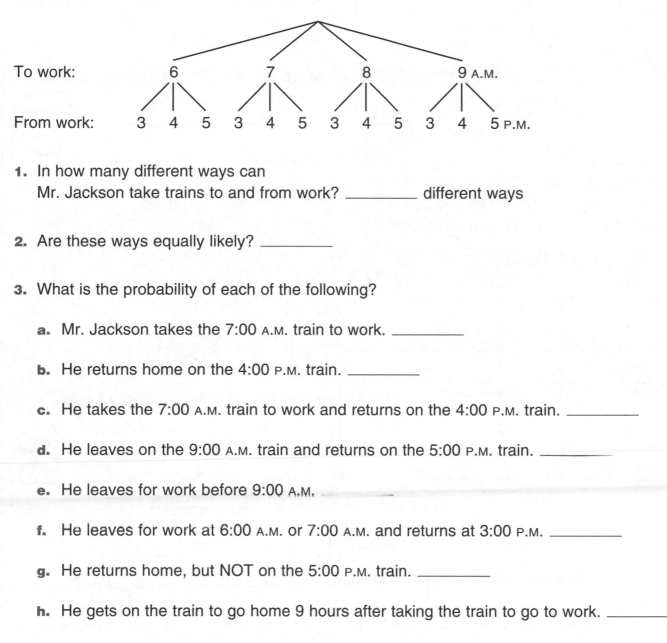

To work: 6   7   8   9 A.M.

From work: 3   4   5   3   4   5   3   4   5   3   4   5 P.M.

1. In how many different ways can
    Mr. Jackson take trains to and from work? _____ different ways

2. Are these ways equally likely? _____

3. What is the probability of each of the following?

    a. Mr. Jackson takes the 7:00 A.M. train to work. _____

    b. He returns home on the 4:00 P.M. train. _____

    c. He takes the 7:00 A.M. train to work and returns on the 4:00 P.M. train. _____

    d. He leaves on the 9:00 A.M. train and returns on the 5:00 P.M. train. _____

    e. He leaves for work before 9:00 A.M. _____

    f. He leaves for work at 6:00 A.M. or 7:00 A.M. and returns at 3:00 P.M. _____

    g. He returns home, but NOT on the 5:00 P.M. train. _____

    h. He gets on the train to go home 9 hours after taking the train to go to work. _____

1. Rename each fraction as a mixed number or a whole number.

a. $\frac{59}{5}$ = _____

b. $\frac{88}{11}$ = _____

c. $\frac{120}{7}$ = _____

d. $\frac{94}{4}$ = _____

e. $\frac{102}{6}$ = _____

SRB
62 63

2. Round each number to the nearest tenth.

a. 50.009 _____

b. 321.65 _____

c. 2.38 _____

d. 0.09 _____

e. 75.993 _____

SRB
45 46

3. The students in Mrs. Dillard's class took a survey of their favorite colors. Complete the table. Then make a circle graph of the data.

Favorite Color	Number of Students	Percent of Class
Red	6	
Blue	10	
Orange	4	
Yellow	2	
Purple	3	
Total		

_____
(title)

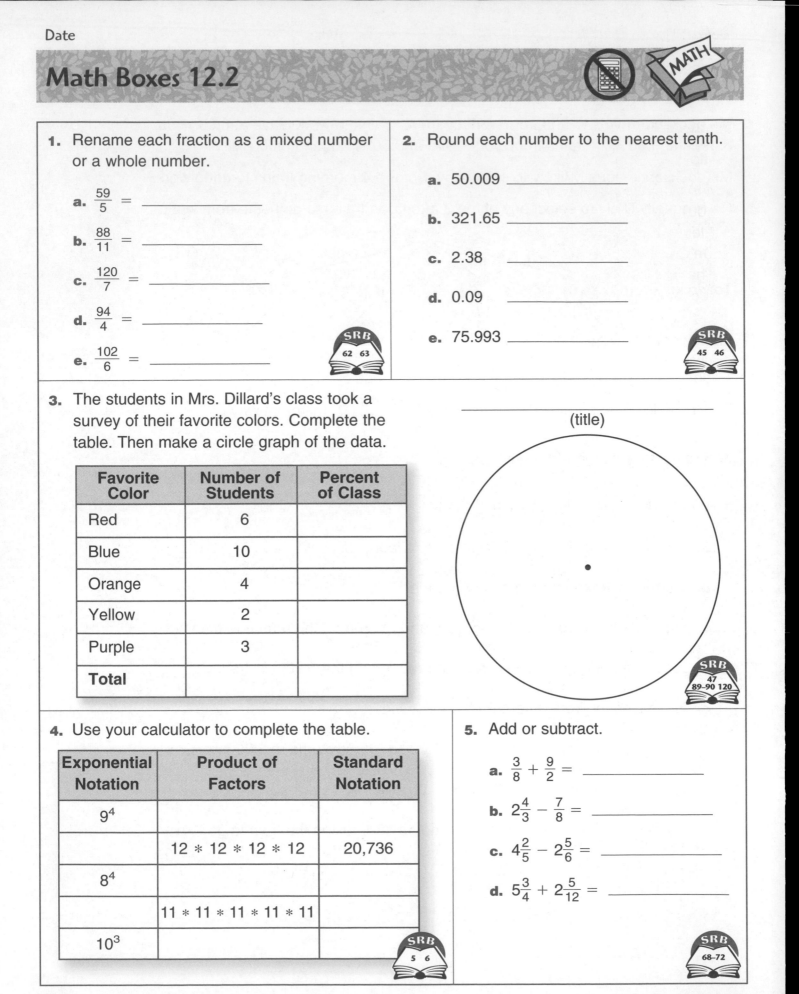

SRB
47
89–90 120

4. Use your calculator to complete the table.

Exponential Notation	Product of Factors	Standard Notation
9^4		
	12 * 12 * 12 * 12	20,736
8^4		
	11 * 11 * 11 * 11 * 11	
10^3		

SRB
5 6

5. Add or subtract.

a. $\frac{3}{8} + \frac{9}{2}$ = _____

b. $2\frac{4}{3} - \frac{7}{8}$ = _____

c. $4\frac{2}{5} - 2\frac{5}{6}$ = _____

d. $5\frac{3}{4} + 2\frac{5}{12}$ = _____

SRB
68–72

Use with Lesson 12.2.

Ratios

Math Message

Ratios can be expressed in many ways. All of the following are statements of ratios:

- It is estimated that by the year 2020 there will be *5 times* as many people 100 years old or older than there were in 1990.
- Elementary school students make up about *14%* of the U.S. population.
- On an average evening, about $\frac{1}{3}$ of the U.S. population watches TV.
- The chances of winning a lottery can be less than *1 in 1 million.*
- A common scale for dollhouses is *1 inch to 12 inches.*

A **ratio** uses division to compare two counts or measures having the same unit. Ratios can be stated or written in a variety of ways. Sometimes a ratio is easier to understand or will make more sense if it is rewritten in another form.

Example In a group of ten students, eight students are right-handed and two are left-handed. The ratio of left-handed students to all students can be expressed in the following ways:

- With words: Two out of the ten students are left-handed.
   Two in ten students are left-handed.
   The ratio of left-handed students to all students is two to ten.
- With a fraction: $\frac{2}{10}$, or $\frac{1}{5}$ of the students are left-handed.
- With a percent: 20% of the students are left-handed.
- With a colon between the two numbers being compared:
   The ratio of left-handed students to all students is 2:10 ("two to ten").

Writing Ratios

Express the ratio of right-handed students to all students in the example above.

1. With words: _____ students are right-handed.

2. With a fraction: _____ of the students are right-handed.

3. With a percent: _____ of the students are right-handed.

4. With a colon: The ratio of right-handed students to all students is _____.

Using Ratios to Examine a Trend

1. a. According to the table on page 314 of the *Student Reference Book,* has the
ratio of farmers to all working people increased or decreased since 1900?

_____

b. Why do you think this has happened? _____

_____

_____

2. a. Has the ratio of engineers to all working people increased or decreased since 1900?

_____

b. Why do you think this has happened? _____

_____

_____

3. a. How has the ratio of clergy to all working people changed since 1900?

_____

b. Why do you think this has happened? _____

_____

_____

Challenge

4. About how many farmers were there

   a. in 1900? _____

   b. in 2000? _____

5. About how many photographers were there

   a. in 1900? _____

   b. in 2000? _____

10 Times

Have you ever heard or used expressions such as "10 times more," "10 times as many," "10 times less," or "$\frac{1}{10}$ as many"? These are **ratio comparisons.** Be sure to use expressions like these with caution. Increasing or reducing something by a factor of 10 makes a big difference!

Scientists call a difference of 10 times a **magnitude,** and they believe that the world as we know it changes drastically when something is increased or decreased by a magnitude.

Example   A person can jog about 5 miles per hour. A car can travel 10 times faster than that, or 50 miles per hour. A plane can travel 10 times faster than that, or 500 miles per hour. Each magnitude increase in travel speed has had a great effect on our lives.

Complete the following table. Then add two of your own events or items to the table.

Event or Item	Current Measure or Count	10 Times More	10 Times Less ($\frac{1}{10}$ as much)
Length of Math Class			
Number of Students in Math Class			
Length of Your Stride			

Math Boxes 12.3

1. Insert parentheses to make each expression true.

   a. $-14 + 36 / 4 - (-2) = -3$

   b. $-14 + 36 / 4 - (-2) = -8$

   c. $15 = (-20) - (-5) + 10 * 3$

   d. $-35 = (-20) - (-5) + 10 * 3$

   e. $8 * 6 - (-24) + 71 = 311$

SRB
203

2. Find the volume of the prism.

| **Volume of a Triangular Box** |
| Volume = Area of the base ∗ height |

9 cm

6 cm

4 cm

Volume: _____ cm^3

SRB
181

3. **a.** Write a 9-digit numeral that has a 4 in the hundred-thousands place, a 6 in the millions place, a 5 in the tens place, a 2 in the hundredths place, a 9 in the thousands place, and a 3 in all of the other places.

      _____ , _____ _____ _____ , _____ _____ _____ . _____ _____

   b. Write this numeral in words.

      _____

      _____

SRB
4
29 30

4. Measure each angle to the nearest degree.

   a.

      D

   ∠*D* measures about _____ °.

   b.

      E

   ∠*E* measures about _____ °.

SRB
188 189

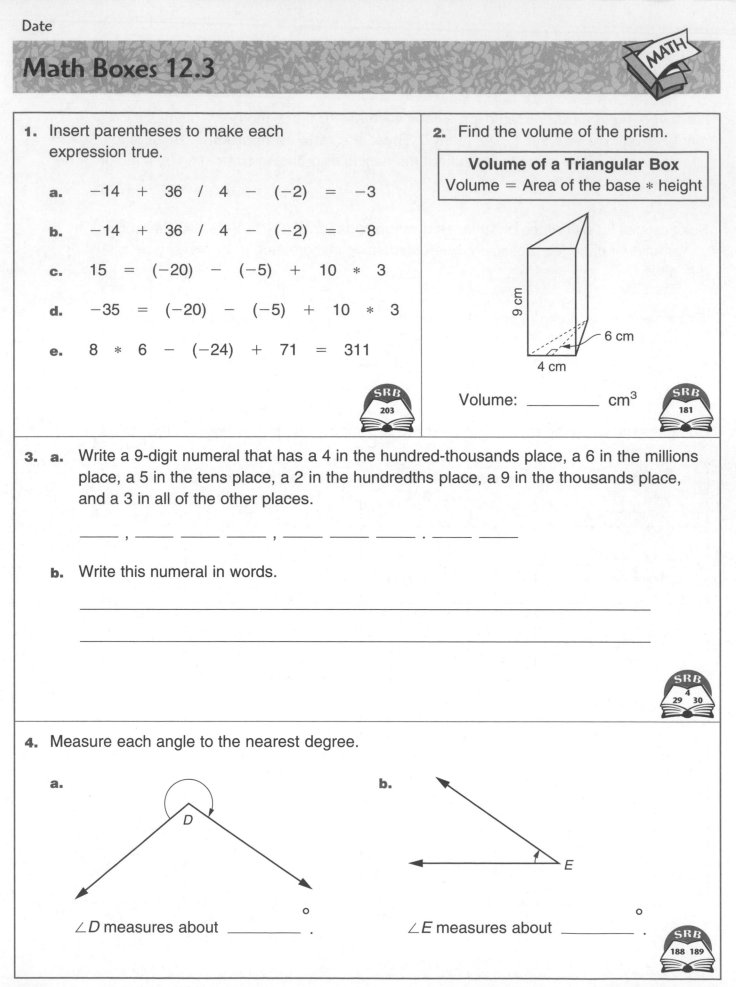

Comparing Parts to Wholes

A **ratio** is a comparison. Some ratios compare part of a collection of things to the total number of things in the collection. For example, the statement "1 out of 6 students in the class is absent" compares the number of students absent to the total number of students in the class. Another way to express this ratio is to say, "For every 6 students enrolled in the class, 1 student is absent" or "$\frac{1}{6}$ of the students in the class are absent."

If you know the total number of students in the class, you can use this ratio to find the number of students who are absent. For example, if there are 12 students in the class, then 2 of the students are absent. If there are 18 students in the class, then 3 students are absent.

If you know the number of students who are absent, you can also use this ratio to find the total number of students in the class. For example, if 5 students are absent, there must be a total of 30 students in the class.

Solve the following ratio problems. Use the square tiles you cut out from *Math Journal 2,* Activity Sheet 8 to help you.

1. Place 28 tiles on your desk so that 1 out of 4 tiles is white and the rest are shaded.

    How many tiles are white? _____   How many tiles are shaded? _____

2. Place 30 tiles on your desk so that 4 out of 5 tiles are white and the rest are shaded.

    How many tiles are white? _____   How many tiles are shaded? _____

3. Place 7 white tiles on your desk. Add some tiles so that 1 out of
    3 tiles is white and the rest are shaded. How many tiles are there in all? _____

4. Place 25 white tiles on your desk. Add some tiles so that 5 out of
    8 tiles are white and the rest are shaded. How many tiles are there in all? _____

5. Take 32 tiles. If 6 out of 8 are white, how many are white? _____

6. Take 15 tiles. If 6 out of 9 are white, how many are white? _____

7. Place 24 tiles on your desk so that 8 are white and the rest are shaded.

    One out of _____ tiles is white.

8. Place 18 tiles on your desk so that 12 are white and the rest are shaded.

    _____ out of 3 tiles are white.

Ratio Number Stories

Use your tiles to model and solve the number stories below.

1. It rained 2 out of 5 days in the month of April. On how many days did it rain that month? _____

2. For every 4 times John was at bat, he got 1 hit. If he got 7 hits, how many times did he bat? _____

3. There are 20 students in Mrs. Kahlid's fifth-grade class. Two out of 8 students have no brothers or sisters. How many students have no brothers or sisters?

   _____

4. Rema eats 2 eggs twice a week. How many eggs will she eat in the month of February? _____

   How many weeks will it take her to eat 32 eggs? _____

5. David took a survey of people's favorite flavors of ice cream. Of the people he surveyed, 2 out of 5 said that they like fudge swirl best, 1 out of 8 chose vanilla, 3 out of 10 chose maple walnut, and the rest chose another flavor.

   a. If 16 people said that fudge swirl is their favorite flavor, how many people took part in David's survey? _____

   b. If 80 people participated in David's survey, how many preferred a flavor that is not fudge swirl, vanilla, or maple walnut? _____

6. Make up your own ratio number story. Ask your partner to solve it.

   _____

   _____

   _____

   _____

   _____

   Answer: _____

Use with Lesson 12.4.

Math Boxes 12.4

1. Martin missed $\frac{1}{8}$ of the 24 shots he took in a basketball game against the Rams.

   a. What fraction of the shots did he make? _____

   b. How many shots did he miss? _____ shots

   c. How many shots did he make? _____ shots

   d. What percent of his shots did he make? _____

2. **a.** Mark and label −1.7, 0.8, −1.3, and 1.9 on the number line.

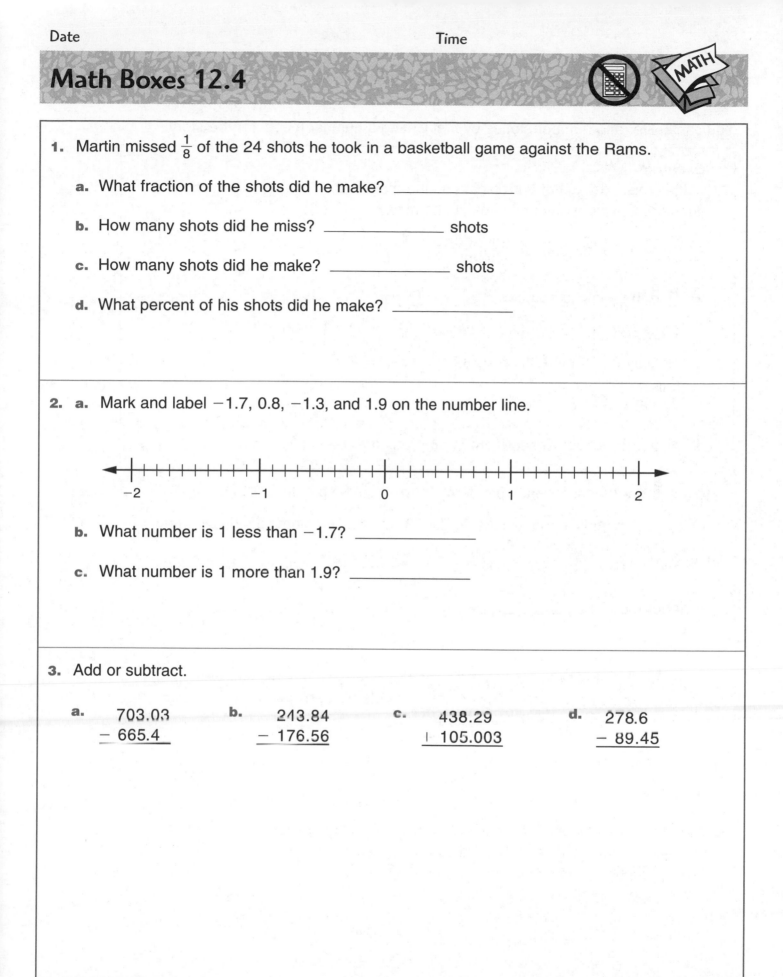

   b. What number is 1 less than −1.7? _____

   c. What number is 1 more than 1.9? _____

3. Add or subtract.

   a. 703.03
         − 665.4

   b. 243.84
         − 176.56

   c. 438.29
         + 105.003

   d. 278.6
         − 89.45

More Ratio Number Stories

You can solve ratio number stories by first writing a number model for the story.

Example

Sidney missed 2 out of 9 problems on the math test. There were 36 problems on the test. How many problems did he miss?

1. Write a number model: $\dfrac{\text{(missed)}}{\text{(total)}} \dfrac{2}{9} = \dfrac{\square}{36}$

2. Find the missing number.

   Think: *9 times what number equals 36?* $9 * 4 = 36$

   Multiply the numerator, 2, by this number: $2 * 4 = 8$

   $\dfrac{\text{(missed)}}{\text{(total)}} \dfrac{2 * 4}{9 * 4} = \dfrac{8}{36}$

3. Answer: Sidney missed 8 out of 36 problems.

Write a number model for each problem. Then solve the problem.

1. Of the 42 animals in the Children's Zoo, 3 out of 7 are mammals. How many mammals are there in the Children's Zoo?

   Number model: _____   Answer: _____

   (unit)

2. Five out of 8 students at Kenwood School play an instrument. There are 224 students at the school. How many students play an instrument?

   Number model: _____   Answer: _____

   (unit)

3. Mr. Lopez sells subscriptions to a magazine. Each subscription costs $18. For each subscription he sells, he earns $8. One week, he sold $198 worth of subscriptions. How much did he earn?

   Number model: _____   Answer: $ _____

More Ratio Number Stories (cont.)

4. Make up a ratio number story. Try to make it a hard one. Ask your partner to solve it.

_____

_____

_____

Answer: _____

Find the missing number.

5. $\dfrac{1}{3} = \dfrac{x}{39}$

    $x =$ _____

6. $\dfrac{3}{4} = \dfrac{21}{y}$

    $y =$ _____

7. $\dfrac{7}{8} = \dfrac{f}{56}$

    $f =$ _____

8. $\dfrac{1}{5} = \dfrac{13}{n}$

    $n =$ _____

9. $\dfrac{5}{6} = \dfrac{m}{42}$

    $m =$ _____

10. $\dfrac{9}{25} = \dfrac{s}{100}$

    $s =$ _____

Challenge

11. There are 48 students in the fifth grade at Robert's school. Three out of 8 fifth graders read two books last month. One out of 3 students read just one book. The rest of the students read no books at all.

    How many books in all did the fifth graders read last month? _____
    (unit)

    Explain what you did to find the answer.

    _____

    _____

    _____

    _____

Volume Review

Area of rectangle: $A = b * h$	Area of circle: $A = \pi * r^2$
Volume of prism or cylinder: $V = B * h$	Circumference of circle: $C = 2 * \pi * r$

1. Find the volume of each cylinder.

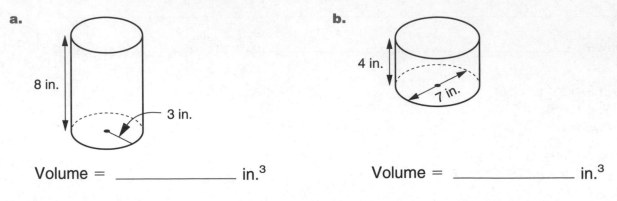

a.

8 in.

3 in.

Volume = _____ in.^{3}

b.

4 in.

7 in.

Volume = _____ in.^{3}

2. Four food containers are pictured below. Find the volume of each. Determine which container has the largest capacity and which has the smallest capacity.

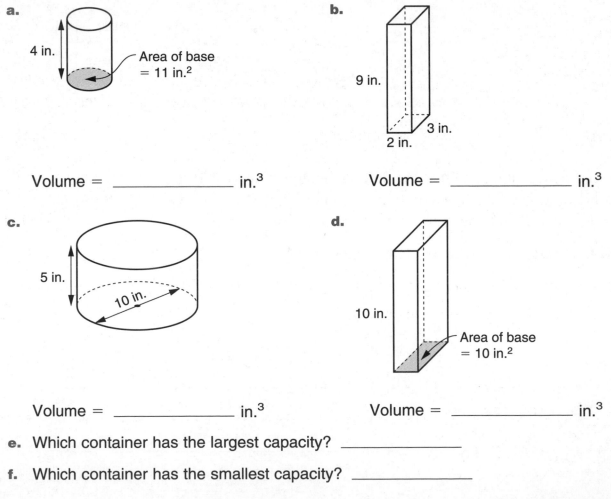

a.

4 in.

Area of base = 11 in.^{2}

Volume = _____ in.^{3}

b.

9 in.

3 in.

2 in.

Volume = _____ in.^{3}

c.

5 in.

10 in.

Volume = _____ in.^{3}

d.

10 in.

Area of base = 10 in.^{2}

Volume = _____ in.^{3}

e. Which container has the largest capacity? _____

f. Which container has the smallest capacity? _____

Use with Lesson 12.5.

Math Boxes 12.5

1. Rename each fraction as a mixed number or a whole number.

a. $\frac{79}{8}$ = _____

b. $\frac{45}{9}$ = _____

c. $\frac{111}{3}$ = _____

d. $\frac{126}{6}$ = _____

e. $\frac{108}{5}$ = _____

2. Round each number to the nearest hundred.

a. 318,495.1 _____

b. 79,002 _____

c. 604.381 _____

d. 13,229 _____

e. 5,098 _____

3. Mrs. Porter's students took a survey of their favorite movie snacks. Complete the table. Then make a circle graph of the data.

Favorite Snack	Number of Students	Percent of Class
Popcorn	11	
Chocolate	5	
Soft drink	6	
Fruit chews	2	
Candy with nuts	1	
Total		

(title)

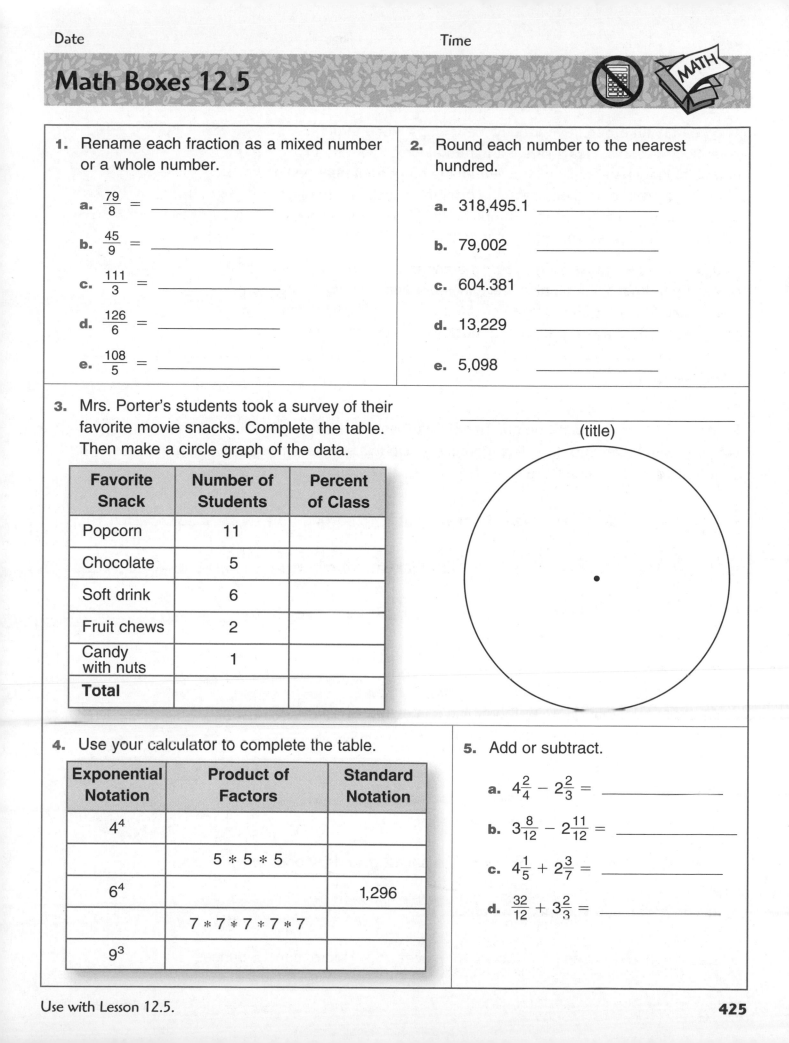

4. Use your calculator to complete the table.

Exponential Notation	Product of Factors	Standard Notation
4^4		
	5 * 5 * 5	
6^4		1,296
	7 * 7 * 7 * 7 * 7	
9^3		

5. Add or subtract.

a. $4\frac{2}{4} - 2\frac{2}{3}$ = _____

b. $3\frac{8}{12} - 2\frac{11}{12}$ = _____

c. $4\frac{1}{5} + 2\frac{3}{7}$ = _____

d. $\frac{32}{12} + 3\frac{2}{3}$ = _____

The Heart

The heart is an organ in your body that pumps blood through your blood vessels. **Heart rate** is the rate at which your heart pumps blood. It is usually expressed as the number of heartbeats per minute. With each heartbeat, the arteries stretch and then go back to their original size. This throbbing of the arteries is called the **pulse.** The **pulse rate** is the same as the heart rate.

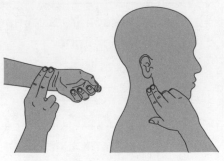

You can feel your pulse along your wrist, near the bone, and below the thumb. You can also feel it in your neck: Run your index and middle fingers from your ear, past the curve of your jaw, and press them into the soft part of your neck just below your jaw.

My Heart Rate

Feel your pulse and count the number of heartbeats in 15 seconds. Your partner can time you with a watch or the classroom clock. Do this several times, until you are sure that your count is accurate.

1. About how many times does your heart beat in 15 seconds? _____

2. At this rate, about how many times would it beat in 1 minute? _____

   in 1 hour? _____

   in 1 day? _____

   in 1 year? _____

3. Your fist and your heart are about the same size. Measure your fist with your ruler. Record the results.

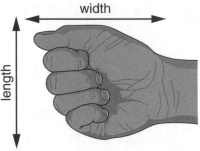

   My heart is about _____ inches wide

   and _____ inches long.

4. A person's heart weighs about 1 ounce per 12 pounds of body weight.

   Circle how much your heart weighs.

   Less than 15 ounces   About 15 ounces   More than 15 ounces

Math Boxes 12.6

1. Insert parentheses to make each expression true.

    a. $-28 + 43 * 2 = 30$

    b. $-19 = 12 / 2 * 6 + (-20)$

    c. $16 = 12 / 2 * 6 + (-20)$

    d. $24 / 6 - (-2) + 5 = 8$

    e. $24 / 6 - (-2) + 5 = 11$

2. Find the volume of the cylinder.

Volume of a Cylinder
Volume = Area of the base * height

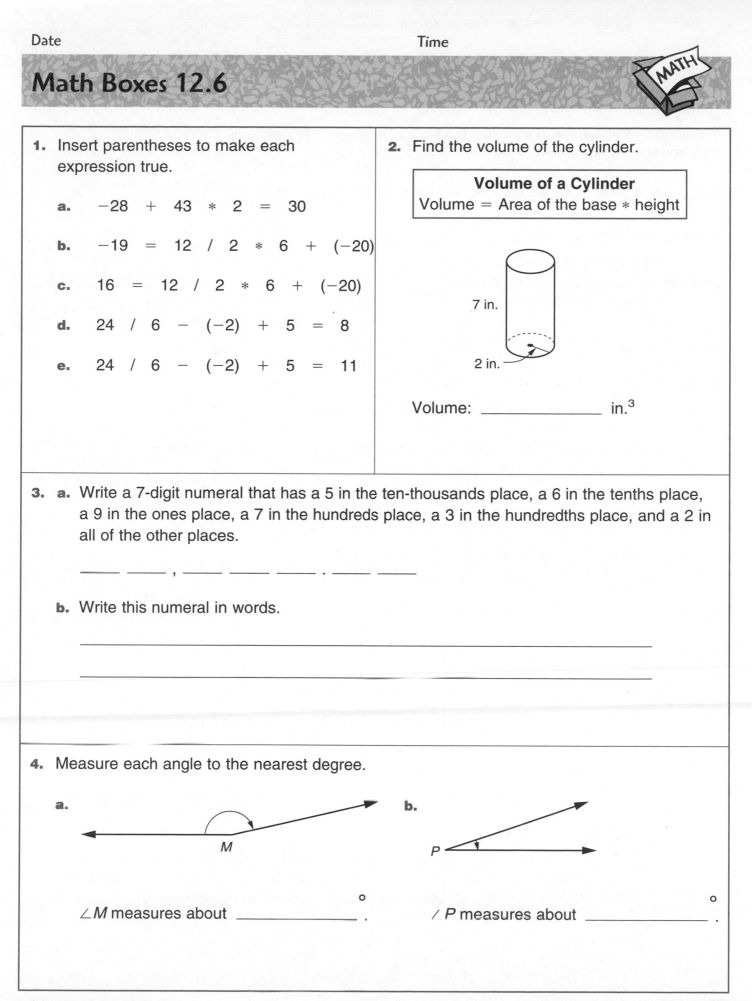

7 in.

2 in.

Volume: _____ in.^{3}

3. a. Write a 7-digit numeral that has a 5 in the ten-thousands place, a 6 in the tenths place, a 9 in the ones place, a 7 in the hundreds place, a 3 in the hundredths place, and a 2 in all of the other places.

    ____ ____ , ____ ____ ____ . ____ ____

    b. Write this numeral in words.

    _____

    _____

4. Measure each angle to the nearest degree.

    a.

    M

    b.

    P

    ∠*M* measures about _____ °.

    ∠*P* measures about _____ °.

Exercise and Your Heart

Exercise increases the rate at which a person's heart beats. Very strenuous exercise can double the heart rate.

Work with a partner to find out how exercise affects your heart rate.

1. Sit quietly for a minute. Then have your partner time you for 15 seconds while you take your pulse. Record the number of heartbeats in the first row of the table at the right.

Step-ups	Heartbeats per 15 Seconds
0	
5	
10	
15	
20	
25	

2. Step up onto and down from a chair 5 times without stopping. As soon as you finish, take your pulse for 15 seconds while your partner times you. Record the number of heartbeats in the second row of the table.

3. Sit quietly. While you are resting, your partner can do 5 step-ups, and you can time your partner.

4. When your pulse is almost back to normal, step up onto and down from the chair 10 times. Record the number of heartbeats in 15 seconds in the third row of the table. Then rest while your partner does 10 step-ups.

5. Repeat for 15, 20, and 25 step-ups.

6. Why is it important that all students step up at the same rate?

_____

_____

_____

Have a Heart

Giraffes do! Their hearts weigh up to 25 pounds and are up to 2 feet across. A giraffe's heart has to work hard to move blood up that neck, which can be 10 to 12 feet long. The average giraffe's blood pressure is three times that of a human's.

Source: Beyond Belief!

My Heart-Rate Profile

1. Make a line graph of the data in your table on journal page 428.

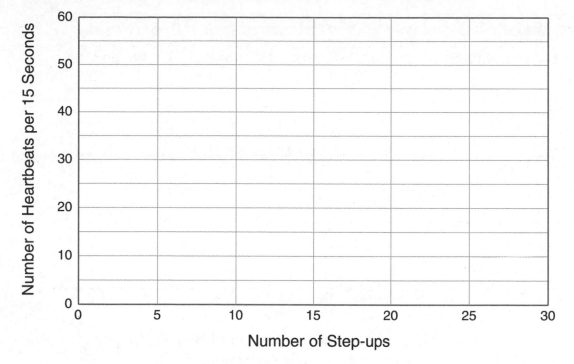

2. Make a prediction: What will your heart rate be if you do 30 step-ups?

    About _____ heartbeats in 15 seconds

3. When you exercise, you must be careful not to put too much stress on your heart. Exercise experts often recommend a "target" heart rate to reach during exercise. The target heart rate varies, depending on a person's age and health, but the following rule is sometimes used.

Target heart rate during exercise:

Subtract your age from 220. Multiply the result by 2. Then divide by 3.

The result is the target number of heartbeats per minute.

a. According to this rule, what is your target heart rate during exercise?

    About _____ heartbeats per minute

b. That's about how many heartbeats in 15 seconds?

    About _____ heartbeats

My Class's Heart-Rate Profile

1. Complete the table.

Class Landmarks: Number of Heartbeats per 15 Seconds				
Number of Step-ups	Maximum	Minimum	Range	Median
0				
5				
10				
15				
20				
25				

2. Make a line graph of the medians on the grid on journal page 429. Use a coloring pencil or crayon. Label this line "Class Profile." Label the other line "My Own Profile."

3. Compare your personal profile to the class profile.

_____

_____

_____

_____

Miles of Blood

There are about 5 million red blood cells, and between 5 thousand and 10 thousand white blood cells, in 1 milliliter of blood from an average man. If all of one man's blood cells were lined up side by side, they would wrap around Earth about seven times.

Source: The Odd Book of Data

Math Boxes 12.7

1. Mark and label each point on the ruler below.

A: $4\frac{1}{4}$ in.   B: $\frac{3}{16}$ in.   C: $2\frac{7}{8}$ in.   D: $1\frac{1}{2}$ in.   E: $3\frac{3}{8}$ in.

```
        |         |         |         |         |
        1         2         3         4         5
  INCHES
```

2. Multiply or divide.

a.   389
    * 20

b.   299
    * 37

c. $9\overline{)243}$

d. $84\overline{)856}$

3. Solve.

a.

One orange weighs

as much as _____ Xs.

One cube weighs

as much as _____ Xs.

b.

One triangle weighs

as much as _____ X.

One paper clip weighs

as much as _____ X.

Review of Ratios

1. What is the ratio of the length of line segment *AB* to the length of line segment *CD*?

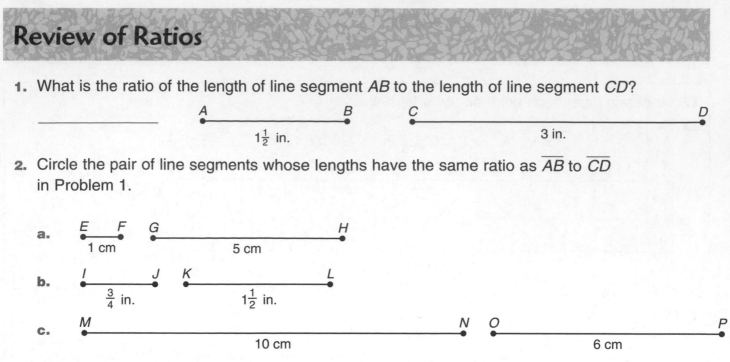

   A ————————— B   C ——————————————————— D
   $1\frac{1}{2}$ in.   3 in.

2. Circle the pair of line segments whose lengths have the same ratio as $\overline{AB}$ to $\overline{CD}$ in Problem 1.

   a.   E · F   G ————————— H
        1 cm   5 cm

   b.   I ——— J   K ————— L
        $\frac{3}{4}$ in. $1\frac{1}{2}$ in.

   c.   M ————————— N   O ——————— P
        10 cm   6 cm

3. There are 13 boys and 15 girls in a group. What fractional part of the group is boys? _____

4. Problem 3 was given to groups of 13-year-olds, 17-year-olds, and adults. The answers and the percent of each group that gave those particular answers are shown in the table below.

Answers	13-Year-Olds	17-Year-Olds	Adults
$\frac{13}{28}$	20%	36%	25%
$\frac{13}{28}$ written as a decimal	0%	0%	1%
$\frac{13}{15}$ or 0.86	17%	17%	15%
$\frac{15}{28}$	2%	2%	3%
Other incorrect answers	44%	29%	35%
Don't know	12%	13%	20%
No answer	5%	3%	1%

   a. What mistake was made by the people who gave the answer $\frac{15}{28}$?

   _____

   _____

   b. What mistake was made by the people who gave the answer $\frac{13}{15}$?

   _____

   _____

Math Boxes 12.8

1. **a.** Measure the radius of the circle in centimeters. _____

    b. Find the area to the nearest cm² and the circumference to the nearest cm.

    > Area = π * radius²
    > Circumference = π * diameter

    The area is about _____.

    The circumference is about _____.

    SRB 171 178

2. To celebrate her birthday, Ms. Hahn decided to give each of the fifth graders one strand of her favorite kind of licorice whip. There are 179 fifth graders. The whips come 15 strands to a package at a cost of $1.19 per package.

    a. How many packages of licorice whips does Ms. Hahn need to buy?

    _____ packages

    b. How much will she spend?

    _____

    SRB 221

3. Complete the table.

Fraction	Decimal	Percent
$\frac{4}{5}$		
		35%
	0.7	
$\frac{8}{20}$		
		87.5%

    SRB 83

4. Complete the table below. Then graph the data and connect the points with line segments.
    Robin runs $\frac{1}{2}$ mile in 4 minutes.
    Rule: $\frac{1}{8}$ * number of minutes = total miles

Number of Minutes	Total Miles
4	$\frac{1}{2}$
8	
	$3\frac{1}{2}$
32	

    Robin's Running Speed

    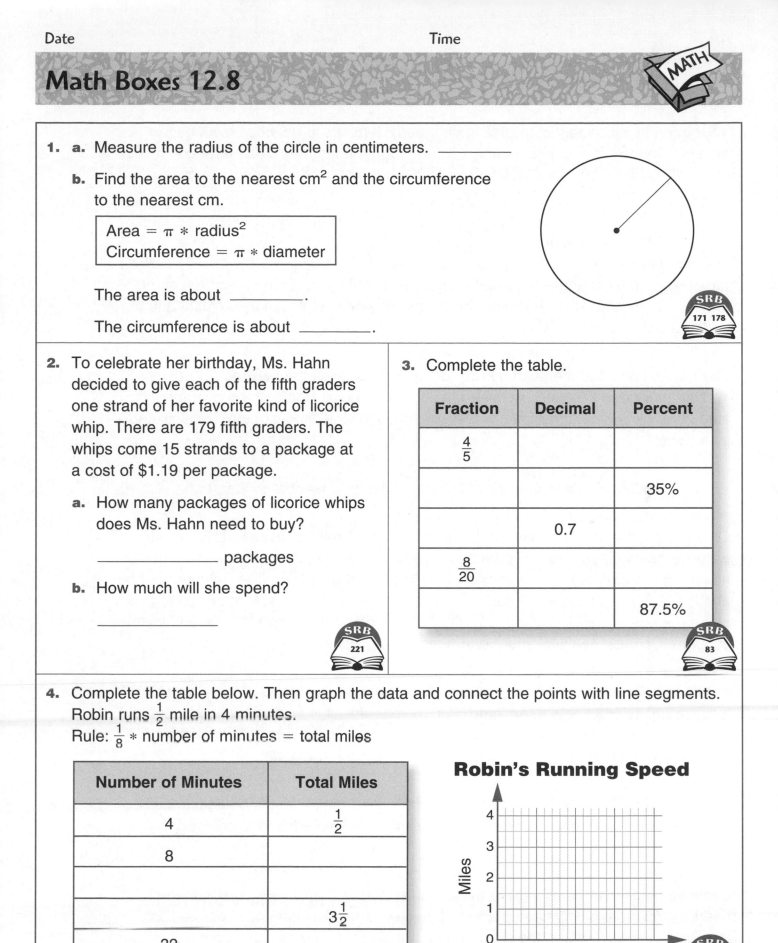

    SRB 217 218

The Heart Pump

Your heart is the strongest muscle in your body. It needs to be, because it never rests. Every day of your life, 24 hours a day, your heart pumps blood throughout your body. The blood carries the **nutrients** and **oxygen** your body needs to function.

You breathe oxygen into your lungs. The oxygen passes from your lungs into your bloodstream. As your heart pumps blood throughout your body, the oxygen is deposited in the cells of your body and is replaced by waste products (mainly **carbon dioxide**). The blood carries the carbon dioxide back to your lungs, which get rid of the carbon dioxide as you exhale. The carbon dioxide is replaced by oxygen, and the cycle begins again.

The amount of blood the heart pumps in 1 minute is called the **cardiac output.** To find your cardiac output, you need to know your **heart rate** and the average amount of blood your heart pumps with each heartbeat. Cardiac output is calculated as follows:

Cardiac output = amount of blood pumped per heartbeat * heart rate

On average, the heart of a fifth grader pumps about 1.6 fluid ounces of blood with each heartbeat. If your heart beats about 90 times per minute, then your heart pumps about 1.6 * 90, or 144 fluid ounces of blood per minute. Your cardiac output would be about 144 fluid ounces, or $1\frac{1}{8}$ gallons of blood per minute. That's about 65 gallons of blood per hour. Imagine having to do this much work, around the clock, every day of your life! Can you see why your heart needs to be very strong?

A person's normal heart rate decreases with age. A newborn's heart rate can be as high as 110 to 160 beats per minute. For 10-year-olds, it is around 90 beats per minute; for adults, it is between 70 and 80 beats per minute. It is not unusual for older people's hearts to beat as few as 50 to 65 times per minute.

Because cardiac output depends on a person's heart rate, it is not the same at all times. The more often the heart beats in 1 minute, the more blood is pumped throughout the body.

Exercise helps your heart grow larger and stronger. The larger and stronger your heart is, the more blood it can pump with each heartbeat. A stronger heart needs fewer heartbeats to pump the same amount of blood. This puts less of a strain on the heart.

The Heart Pump (cont.)

Pretend that your heart has been pumping the same amount of blood all of your life so far—about 65 gallons of blood per hour.

1. a. At that rate, about how many gallons of blood would your heart pump per day?

About _____ gallons

b. About how many gallons per year? About _____ gallons

2. At that rate, about how many gallons would it have pumped from the time you were born to your last birthday? About _____ gallons

3. Both heart rate and cardiac output increase with exercise. Look at the table on journal page 428. Find the number of heartbeats in 15 seconds when you are at rest and the number of heartbeats after 25 step-ups. Record them below.

a. Heartbeats in 15 seconds at rest: _____

b. Heartbeats in 15 seconds after 25 step-ups: _____

Now figure out how many heartbeats in 1 minute.

c. Heartbeats in 1 minute at rest: _____

d. Heartbeats in 1 minute after 25 step-ups: _____

4. If your heart pumps about 1.6 fluid ounces of blood per heartbeat, about how much blood does it pump in 1 minute when you are at rest?

About _____ fl oz

5. A gallon is equal to 128 fluid ounces. About how many gallons of blood does your heart pump in 1 minute when you are at rest?

About _____ gallon(s)

6. a. Use your answer to Problem 5 above to find about how many fluid ounces of blood your heart would pump in 1 minute after 25 step-ups.

About _____ fl oz

b. About how many gallons? About _____ gallon(s)

Use with Lesson 12.8.

American Tour: End-of-Year Projects

Work with a partner or in a small group on one or more of the following projects, or think up a project of your own. Each project has four steps.

Step 1 Plan and Do Research. Use the American Tour section of the *Student Reference Book* and other reference sources such as encyclopedias, almanacs, and the Internet to locate necessary and helpful data. Decide which information to use.

Step 2 Analyze Data. In order to complete the project, you will need to analyze and possibly transform the data you find in your sources.

Step 3 Record and Display Your Findings. Write a journal; make charts, graphs, tables, and other displays to record and show what you have found.

Step 4 Present Your Results. Report your findings to your classmates in clear and interesting ways.

Project 1: "Most" State, "Least" State, My State

Look up a variety of population and environmental statistics in the American Tour and in other sources. Create a display that shows which state has the most or the highest number, which has the least or lowest number, and the number for your state (if it is different). Then write a sentence or two for each comparison that describes how your state compares to the "most" and "least" states.

For example, if you live in Connecticut, you might make a comparison like the one below:

Population in 2000		
Most	California	32,521,000 people
Least	Wyoming	525,000 people
My State	Connecticut	3,284,000 people

My state's population is about $\frac{1}{10}$ the population of California, but about 6 times the population of Wyoming.

This is just one way you could make the comparison. There are many others. Find ways to make interesting and informative comparisons.

American Tour: End-of-Year Projects (cont.)

Project 2: Then and Now

The American Tour section of the *Student Reference Book* contains information about the United States during your lifetime and approximately 100 and 200 years ago. Use some of this information, as well as information from other sources, to create a series of bar graphs that compare the United States of your lifetime to the United States of approximately 100 and 200 years ago. For each graph, write a newspaper headline that describes an interesting pattern or fact shown by the graph.

For example, one of the bar graphs might compare the percent of the population living in urban areas in 1790, 1900, and 2000. It might look like the graph at the right.

Percent of U.S. Population Living in Urban Areas Increases 15-Fold in 200 Years!

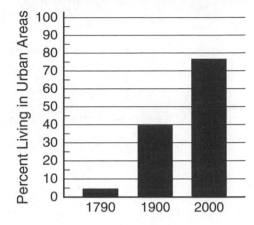

Some hints to keep in mind:

- Sometimes there will be no data for approximately 200 years ago. If this is the case, then compare data for approximately 100 years ago with the present.

- The dates do not have to be the same for each bar graph. Just make sure to note on each graph which years you are comparing.

- Clearly label your graphs. Give them titles, and indicate which counts, measures, or percents you are comparing and the years for which you have data.

Project 3: State Almanac

Use the American Tour and other sources to create a State Almanac of interesting facts and features about your state (or another state). You might include the following information:

- the year your state became a state

- the number of Native Americans who lived in your state in 2000

- the number of times greater your state's population was in 2000 than in 1900

Illustrate the State Almanac with graphs, pictures, and other displays that highlight special features of your state.

American Tour: End-of-Year Projects (cont.)

Project 4: A Westward Journey

Use the information in the American Tour section of the *Student Reference Book* to help you write a journal that describes a trip across the country in 1840.

Begin the trip at a city on the east coast. From there, travel to St. Louis. Make part of this journey by foot, part by horseback, and part by stagecoach.

From St. Louis, take the trail of your choice west. Assume there is a road from St. Louis to Independence along the Missouri River.

Make approximately half of the journey from St. Louis to a city on the west coast by stagecoach and half by wagon train.

Find the number of days each part of the trip will take and the total traveling time from coast to coast.

You will need to make other decisions. How many hours per day could you travel by the various means of transportation? Do you need to rest along the way? Use your imagination.

For travel between cities in the northeast and St. Louis, you can use the highway map on page 346 as a rough guide to distances between cities. For travel west of St. Louis, use the map and scale on page 312.

You might, for example, begin as follows:

June 1   We departed Boston by stagecoach. Our destination was New York.

June 3   We arrived in New York. The journey from Boston took 2 days. The stagecoach traveled about 12 hours a day, covering around 100 miles each day. We were exhausted and so were the horses!

June 4   We left for Pittsburgh via Philadelphia and Lancaster, traveling by horseback.

June 12   The 400-mile horseback journey to Pittsburgh took 8 days. We covered about 65 miles per day, but we could not travel for two days due to driving rainstorms that washed out the road.

                                                           Use with Lesson 12.9.

Math Boxes 12.9

1. Mark and label each point on the ruler below.

   A: $3\frac{3}{8}$ "   B: $1\frac{5}{16}$ "   C: $\frac{15}{16}$ "   D: $4\frac{5}{8}$ "   E: $2\frac{3}{4}$ "

   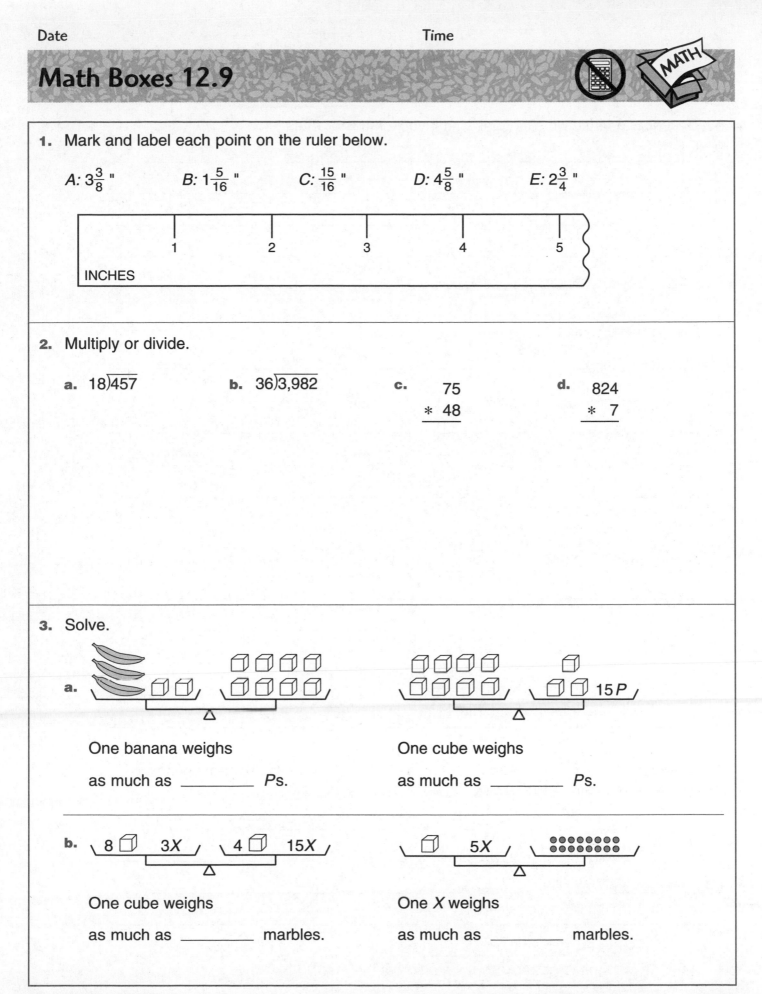

   INCHES

2. Multiply or divide.

   a. $18\overline{)457}$ b. $36\overline{)3,982}$

   c.   75   d.   824
      $* 48$ $* 7$

3. Solve.

   a.

   One banana weighs   One cube weighs

   as much as _____ Ps.   as much as _____ Ps.

   b. 8 ☐ 3X   4 ☐ 15X   ☐ 5X   •••••••••

   One cube weighs   One X weighs

   as much as _____ marbles.   as much as _____ marbles.

Time to Reflect

1. Describe at least two situations that involve ratios or rates.

   _____

   _____

   _____

   _____

   _____

2. Explain what a ratio is.

   _____

   _____

   _____

   _____

   _____

3. Describe something from this unit that you think next year's fifth graders might find confusing.

   _____

   _____

   _____

   _____

   _____

4. Look back through your journal. List at least one topic in this unit that you think you will find useful. Explain why.

   _____

   _____

   _____

   _____

Use with Lesson 12.10.

Math Boxes 12.10

1. **a.** Draw a circle with a radius of 2.5 centimeters.

   b. What is the area of this circle to the nearest centimeter?

   $$\boxed{\text{Area} = \pi * \text{radius}^2}$$

   About _____
   (unit)

2. Cherie lives on a chicken farm. Every morning she collects the eggs from the hen house. She collected 162 eggs on Monday, 104 eggs on Tuesday, and 157 eggs on Wednesday. She packed them into cartons, each containing 12 eggs. How many egg cartons can she fill completely?

   _____ egg cartons

3. Complete the table.

Fraction	Decimal	Percent
	0.18	
		37.5%
$\frac{45}{50}$		
$\frac{16}{25}$		
	0.88	

4. Complete the table. Then graph the data and connect the points with line segments. David rides his bike at a speed of about 12 miles per hour.

Rule: 12 * number of hours = total miles

Number of Hours	Total Miles
1	12
2	
	36
	42
5	

David's Bike Speed

Reference

Equivalent Fractions, Decimals, and Percents

															Decimal	Percent
$\frac{1}{2}$	$\frac{2}{4}$	$\frac{3}{6}$	$\frac{4}{8}$	$\frac{5}{10}$	$\frac{6}{12}$	$\frac{7}{14}$	$\frac{8}{16}$	$\frac{9}{18}$	$\frac{10}{20}$	$\frac{11}{22}$	$\frac{12}{24}$	$\frac{13}{26}$	$\frac{14}{28}$	$\frac{15}{30}$	0.5	50%
$\frac{1}{3}$	$\frac{2}{6}$	$\frac{3}{9}$	$\frac{4}{12}$	$\frac{5}{15}$	$\frac{6}{18}$	$\frac{7}{21}$	$\frac{8}{24}$	$\frac{9}{27}$	$\frac{10}{30}$	$\frac{11}{33}$	$\frac{12}{36}$	$\frac{13}{39}$	$\frac{14}{42}$	$\frac{15}{45}$	$0.\overline{3}$	$33\frac{1}{3}\%$
$\frac{2}{3}$	$\frac{4}{6}$	$\frac{6}{9}$	$\frac{8}{12}$	$\frac{10}{15}$	$\frac{12}{18}$	$\frac{14}{21}$	$\frac{16}{24}$	$\frac{18}{27}$	$\frac{20}{30}$	$\frac{22}{33}$	$\frac{24}{36}$	$\frac{26}{39}$	$\frac{28}{42}$	$\frac{30}{45}$	$0.\overline{6}$	$66\frac{2}{3}\%$
$\frac{1}{4}$	$\frac{2}{8}$	$\frac{3}{12}$	$\frac{4}{16}$	$\frac{5}{20}$	$\frac{6}{24}$	$\frac{7}{28}$	$\frac{8}{32}$	$\frac{9}{36}$	$\frac{10}{40}$	$\frac{11}{44}$	$\frac{12}{48}$	$\frac{13}{52}$	$\frac{14}{56}$	$\frac{15}{60}$	0.25	25%
$\frac{3}{4}$	$\frac{6}{8}$	$\frac{9}{12}$	$\frac{12}{16}$	$\frac{15}{20}$	$\frac{18}{24}$	$\frac{21}{28}$	$\frac{24}{32}$	$\frac{27}{36}$	$\frac{30}{40}$	$\frac{33}{44}$	$\frac{36}{48}$	$\frac{39}{52}$	$\frac{42}{56}$	$\frac{45}{60}$	0.75	75%
$\frac{1}{5}$	$\frac{2}{10}$	$\frac{3}{15}$	$\frac{4}{20}$	$\frac{5}{25}$	$\frac{6}{30}$	$\frac{7}{35}$	$\frac{8}{40}$	$\frac{9}{45}$	$\frac{10}{50}$	$\frac{11}{55}$	$\frac{12}{60}$	$\frac{13}{65}$	$\frac{14}{70}$	$\frac{15}{75}$	0.2	20%
$\frac{2}{5}$	$\frac{4}{10}$	$\frac{6}{15}$	$\frac{8}{20}$	$\frac{10}{25}$	$\frac{12}{30}$	$\frac{14}{35}$	$\frac{16}{40}$	$\frac{18}{45}$	$\frac{20}{50}$	$\frac{22}{55}$	$\frac{24}{60}$	$\frac{26}{65}$	$\frac{28}{70}$	$\frac{30}{75}$	0.4	40%
$\frac{3}{5}$	$\frac{6}{10}$	$\frac{9}{15}$	$\frac{12}{20}$	$\frac{15}{25}$	$\frac{18}{30}$	$\frac{21}{35}$	$\frac{24}{40}$	$\frac{27}{45}$	$\frac{30}{50}$	$\frac{33}{55}$	$\frac{36}{60}$	$\frac{39}{65}$	$\frac{42}{70}$	$\frac{45}{75}$	0.6	60%
$\frac{4}{5}$	$\frac{8}{10}$	$\frac{12}{15}$	$\frac{16}{20}$	$\frac{20}{25}$	$\frac{24}{30}$	$\frac{28}{35}$	$\frac{32}{40}$	$\frac{36}{45}$	$\frac{40}{50}$	$\frac{44}{55}$	$\frac{48}{60}$	$\frac{52}{65}$	$\frac{56}{70}$	$\frac{60}{75}$	0.8	80%
$\frac{1}{6}$	$\frac{2}{12}$	$\frac{3}{18}$	$\frac{4}{24}$	$\frac{5}{30}$	$\frac{6}{36}$	$\frac{7}{42}$	$\frac{8}{48}$	$\frac{9}{54}$	$\frac{10}{60}$	$\frac{11}{66}$	$\frac{12}{72}$	$\frac{13}{78}$	$\frac{14}{84}$	$\frac{15}{90}$	$0.1\overline{6}$	$16\frac{2}{3}\%$
$\frac{5}{6}$	$\frac{10}{12}$	$\frac{15}{18}$	$\frac{20}{24}$	$\frac{25}{30}$	$\frac{30}{36}$	$\frac{35}{42}$	$\frac{40}{48}$	$\frac{45}{54}$	$\frac{50}{60}$	$\frac{55}{66}$	$\frac{60}{72}$	$\frac{65}{78}$	$\frac{70}{84}$	$\frac{75}{90}$	$0.8\overline{3}$	$83\frac{1}{3}\%$
$\frac{1}{7}$	$\frac{2}{14}$	$\frac{3}{21}$	$\frac{4}{28}$	$\frac{5}{35}$	$\frac{6}{42}$	$\frac{7}{49}$	$\frac{8}{56}$	$\frac{9}{63}$	$\frac{10}{70}$	$\frac{11}{77}$	$\frac{12}{84}$	$\frac{13}{91}$	$\frac{14}{98}$	$\frac{15}{105}$	0.143	14.3%
$\frac{2}{7}$	$\frac{4}{14}$	$\frac{6}{21}$	$\frac{8}{28}$	$\frac{10}{35}$	$\frac{12}{42}$	$\frac{14}{49}$	$\frac{16}{56}$	$\frac{18}{63}$	$\frac{20}{70}$	$\frac{22}{77}$	$\frac{24}{84}$	$\frac{26}{91}$	$\frac{28}{98}$	$\frac{30}{105}$	0.286	28.6%
$\frac{3}{7}$	$\frac{6}{14}$	$\frac{9}{21}$	$\frac{12}{28}$	$\frac{15}{35}$	$\frac{18}{42}$	$\frac{21}{49}$	$\frac{24}{56}$	$\frac{27}{63}$	$\frac{30}{70}$	$\frac{33}{77}$	$\frac{36}{84}$	$\frac{39}{91}$	$\frac{42}{98}$	$\frac{45}{105}$	0.429	42.9%
$\frac{4}{7}$	$\frac{8}{14}$	$\frac{12}{21}$	$\frac{16}{28}$	$\frac{20}{35}$	$\frac{24}{42}$	$\frac{28}{49}$	$\frac{32}{56}$	$\frac{36}{63}$	$\frac{40}{70}$	$\frac{44}{77}$	$\frac{48}{84}$	$\frac{52}{91}$	$\frac{56}{98}$	$\frac{60}{105}$	0.571	57.1%
$\frac{5}{7}$	$\frac{10}{14}$	$\frac{15}{21}$	$\frac{20}{28}$	$\frac{25}{35}$	$\frac{30}{42}$	$\frac{35}{49}$	$\frac{40}{56}$	$\frac{45}{63}$	$\frac{50}{70}$	$\frac{55}{77}$	$\frac{60}{84}$	$\frac{65}{91}$	$\frac{70}{98}$	$\frac{75}{105}$	0.714	71.4%
$\frac{6}{7}$	$\frac{12}{14}$	$\frac{18}{21}$	$\frac{24}{28}$	$\frac{30}{35}$	$\frac{36}{42}$	$\frac{42}{49}$	$\frac{48}{56}$	$\frac{54}{63}$	$\frac{60}{70}$	$\frac{66}{77}$	$\frac{72}{84}$	$\frac{78}{91}$	$\frac{84}{98}$	$\frac{90}{105}$	0.857	85.7%
$\frac{1}{8}$	$\frac{2}{16}$	$\frac{3}{24}$	$\frac{4}{32}$	$\frac{5}{40}$	$\frac{6}{48}$	$\frac{7}{56}$	$\frac{8}{64}$	$\frac{9}{72}$	$\frac{10}{80}$	$\frac{11}{88}$	$\frac{12}{96}$	$\frac{13}{104}$	$\frac{14}{112}$	$\frac{15}{120}$	0.125	$12\frac{1}{2}\%$
$\frac{3}{8}$	$\frac{6}{16}$	$\frac{9}{24}$	$\frac{12}{32}$	$\frac{15}{40}$	$\frac{18}{48}$	$\frac{21}{56}$	$\frac{24}{64}$	$\frac{27}{72}$	$\frac{30}{80}$	$\frac{33}{88}$	$\frac{36}{96}$	$\frac{39}{104}$	$\frac{42}{112}$	$\frac{45}{120}$	0.375	$37\frac{1}{2}\%$
$\frac{5}{8}$	$\frac{10}{16}$	$\frac{15}{24}$	$\frac{20}{32}$	$\frac{25}{40}$	$\frac{30}{48}$	$\frac{35}{56}$	$\frac{40}{64}$	$\frac{45}{72}$	$\frac{50}{80}$	$\frac{55}{88}$	$\frac{60}{96}$	$\frac{65}{104}$	$\frac{70}{112}$	$\frac{75}{120}$	0.625	$62\frac{1}{2}\%$
$\frac{7}{8}$	$\frac{14}{16}$	$\frac{21}{24}$	$\frac{28}{32}$	$\frac{35}{40}$	$\frac{42}{48}$	$\frac{49}{56}$	$\frac{56}{64}$	$\frac{63}{72}$	$\frac{70}{80}$	$\frac{77}{88}$	$\frac{84}{96}$	$\frac{91}{104}$	$\frac{98}{112}$	$\frac{105}{120}$	0.875	$87\frac{1}{2}\%$
$\frac{1}{9}$	$\frac{2}{18}$	$\frac{3}{27}$	$\frac{4}{36}$	$\frac{5}{45}$	$\frac{6}{54}$	$\frac{7}{63}$	$\frac{8}{72}$	$\frac{9}{81}$	$\frac{10}{90}$	$\frac{11}{99}$	$\frac{12}{108}$	$\frac{13}{117}$	$\frac{14}{126}$	$\frac{15}{135}$	$0.\overline{1}$	$11\frac{1}{9}\%$
$\frac{2}{9}$	$\frac{4}{18}$	$\frac{6}{27}$	$\frac{8}{36}$	$\frac{10}{45}$	$\frac{12}{54}$	$\frac{14}{63}$	$\frac{16}{72}$	$\frac{18}{81}$	$\frac{20}{90}$	$\frac{22}{99}$	$\frac{24}{108}$	$\frac{26}{117}$	$\frac{28}{126}$	$\frac{30}{135}$	$0.\overline{2}$	$22\frac{2}{9}\%$
$\frac{4}{9}$	$\frac{8}{18}$	$\frac{12}{27}$	$\frac{16}{36}$	$\frac{20}{45}$	$\frac{24}{54}$	$\frac{28}{63}$	$\frac{32}{72}$	$\frac{36}{81}$	$\frac{40}{90}$	$\frac{44}{99}$	$\frac{48}{108}$	$\frac{52}{117}$	$\frac{56}{126}$	$\frac{60}{135}$	$0.\overline{4}$	$44\frac{4}{9}\%$
$\frac{5}{9}$	$\frac{10}{18}$	$\frac{15}{27}$	$\frac{20}{36}$	$\frac{25}{45}$	$\frac{30}{54}$	$\frac{35}{63}$	$\frac{40}{72}$	$\frac{45}{81}$	$\frac{50}{90}$	$\frac{55}{99}$	$\frac{60}{108}$	$\frac{65}{117}$	$\frac{70}{126}$	$\frac{75}{135}$	$0.\overline{5}$	$55\frac{5}{9}\%$
$\frac{7}{9}$	$\frac{14}{18}$	$\frac{21}{27}$	$\frac{28}{36}$	$\frac{35}{45}$	$\frac{42}{54}$	$\frac{49}{63}$	$\frac{56}{72}$	$\frac{63}{81}$	$\frac{70}{90}$	$\frac{77}{99}$	$\frac{84}{108}$	$\frac{91}{117}$	$\frac{98}{126}$	$\frac{105}{135}$	$0.\overline{7}$	$77\frac{7}{9}\%$
$\frac{8}{9}$	$\frac{16}{18}$	$\frac{24}{27}$	$\frac{32}{36}$	$\frac{40}{45}$	$\frac{48}{54}$	$\frac{56}{63}$	$\frac{64}{72}$	$\frac{72}{81}$	$\frac{80}{90}$	$\frac{88}{99}$	$\frac{96}{108}$	$\frac{104}{117}$	$\frac{112}{126}$	$\frac{120}{135}$	$0.\overline{8}$	$88\frac{8}{9}\%$

Note: The decimals for sevenths have been rounded to the nearest thousandth.

Reference

Metric System

Units of Length

1 kilometer (km)	=	1000 meters (m)
1 meter	=	10 decimeters (dm)
	=	100 centimeters (cm)
	=	1000 millimeters (mm)
1 decimeter	=	10 centimeters
1 centimeter	=	10 millimeters

Units of Area

1 square meter (m^2)	=	100 square decimeters (dm^2)
	=	10,000 square centimeters (cm^2)
1 square decimeter	=	100 square centimeters
1 are (a)	=	100 square meters
1 hectare (ha)	=	100 ares
1 square kilometer (km^2)	=	100 hectares

Units of Volume

1 cubic meter (m^3)	=	1000 cubic decimeters (dm^3)
	=	1,000,000 cubic centimeters (cm^3)
1 cubic decimeter	=	1000 cubic centimeters

Units of Capacity

1 kiloliter (kL)	=	1000 liters (L)
1 liter	=	1000 milliliters (mL)

Units of Mass

1 metric ton (t)	=	1000 kilograms (kg)
1 kilogram	=	1000 grams (g)
1 gram	=	1000 milligrams (mg)

Units of Time

1 century	=	100 years
1 decade	=	10 years
1 year (yr)	=	12 months
	=	52 weeks (plus one or two days)
	=	365 days (366 days in a leap year)
1 month (mo)	=	28, 29, 30, or 31 days
1 week (wk)	=	7 days
1 day (d)	=	24 hours
1 hour (hr)	=	60 minutes
1 minute (min)	=	60 seconds (sec)

U.S. Customary System

Units of Length

1 mile (mi)	=	1760 yards (yd)
	=	5280 feet (ft)
1 yard	=	3 feet
	=	36 inches (in.)
1 foot	=	12 inches

Units of Area

1 square yard (yd^2)	=	9 square feet (ft^2)
	=	1296 square inches ($in.^2$)
1 square foot	=	144 square inches
1 acre	=	43,560 square feet
1 square mile (mi^2)	=	640 acres

Units of Volume

1 cubic yard (yd^3)	=	27 cubic feet (ft^3)
1 cubic foot	=	1728 cubic inches ($in.^3$)

Units of Capacity

1 gallon (gal)	=	4 quarts (qt)
1 quart	=	2 pints (pt)
1 pint	=	2 cups (c)
1 cup	=	8 fluid ounces (fl oz)
1 fluid ounce	=	2 tablespoons (tbs)
1 tablespoon	=	3 teaspoons (tsp)

Units of Weight

1 ton (T)	=	2000 pounds (lb)
1 pound	=	16 ounces (oz)

System Equivalents

1 inch is about 2.5 cm (2.54)

1 kilometer is about 0.6 mile (0.621)

1 mile is about 1.6 kilometers (1.609)

1 meter is about 39 inches (39.37)

1 liter is about 1.1 quarts (1.057)

1 ounce is about 28 grams (28.350)

1 kilogram is about 2.2 pounds (2.205)

1 hectare is about 2.5 acres (2.47)

Rules for Order of Operations

1. Do operations within parentheses or other grouping symbols before doing anything else.
2. Calculate all powers.
3. Do multiplications or divisions in order, from left to right.
4. Then do additions or subtractions in order, from left to right.

Reference

Place-Value Chart

trillions	100B	10B	billions	100M	10M	millions	hundred-thousands	ten-thousands	thousands	hundreds	tens	ones	.	tenths	hundredths	thousandths
1000 billions			1000 millions			1,000,000s	100,000s	10,000s	1000s	100s	10s	1s	.	0.1s	0.01s	0.001s
10^{12}	10^{11}	10^{10}	10^9	10^8	10^7	10^6	10^5	10^4	10^3	10^2	10^1	10^0	.	10^{-1}	10^{-2}	10^{-3}

Probability Meter

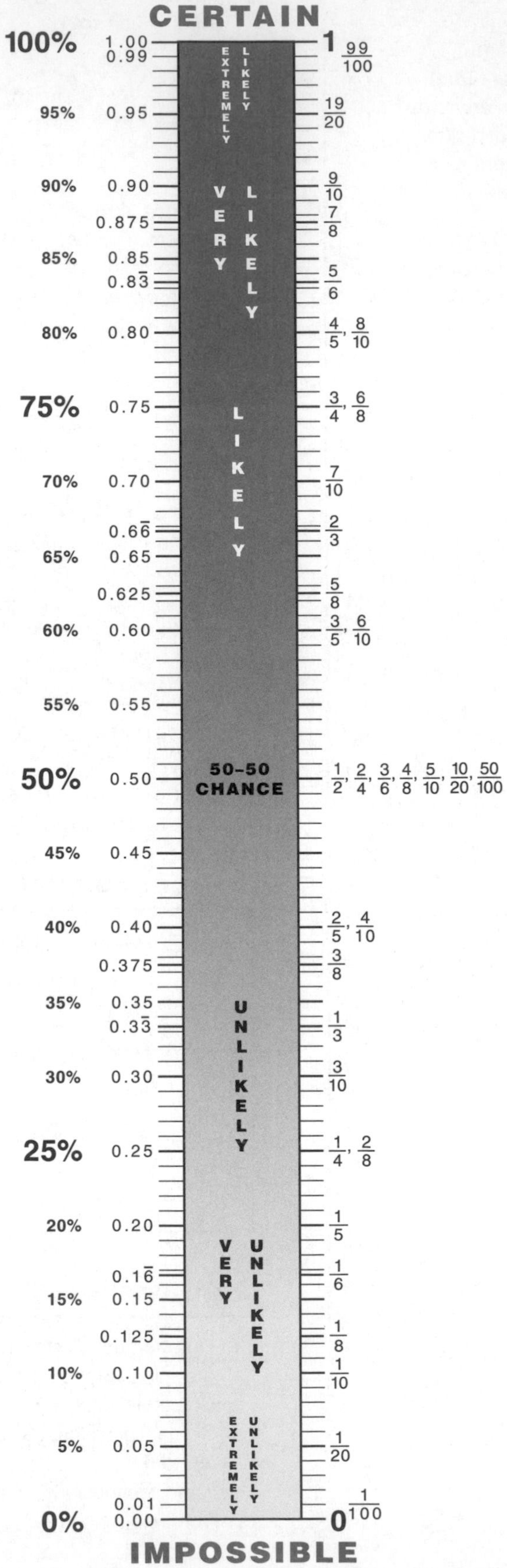

CERTAIN

%	Decimal	Fraction
100%	1.00 / 0.99	1 , $\frac{99}{100}$
95%	0.95	$\frac{19}{20}$
90%	0.90	$\frac{9}{10}$
	0.875	$\frac{7}{8}$
85%	0.85 / 0.8$\overline{3}$	$\frac{5}{6}$
80%	0.80	$\frac{4}{5}$, $\frac{8}{10}$
75%	0.75	$\frac{3}{4}$, $\frac{6}{8}$
70%	0.70	$\frac{7}{10}$
	0.6$\overline{6}$	$\frac{2}{3}$
65%	0.65	
	0.625	$\frac{5}{8}$
60%	0.60	$\frac{3}{5}$, $\frac{6}{10}$
55%	0.55	
50%	0.50	$\frac{1}{2}$, $\frac{2}{4}$, $\frac{3}{6}$, $\frac{4}{8}$, $\frac{5}{10}$, $\frac{10}{20}$, $\frac{50}{100}$
45%	0.45	
40%	0.40	$\frac{2}{5}$, $\frac{4}{10}$
	0.375	$\frac{3}{8}$
35%	0.35 / 0.3$\overline{3}$	$\frac{1}{3}$
30%	0.30	$\frac{3}{10}$
25%	0.25	$\frac{1}{4}$, $\frac{2}{8}$
20%	0.20	$\frac{1}{5}$
	0.1$\overline{6}$	$\frac{1}{6}$
15%	0.15	
	0.125	$\frac{1}{8}$
10%	0.10	$\frac{1}{10}$
5%	0.05	$\frac{1}{20}$
	0.01	
0%	0.00	0 , $\frac{1}{100}$

Labels on meter (top to bottom): EXTREMELY LIKELY, VERY LIKELY, LIKELY, 50–50 CHANCE, UNLIKELY, VERY UNLIKELY, EXTREMELY UNLIKELY

IMPOSSIBLE

Symbols

$+$	plus or positive
$-$	minus or negative
$*$, $\times$	multiplied by
$\div$, $/$	divided by
$=$	is equal to
$\neq$	is not equal to
$<$	is less than
$>$	is greater than
$\leq$	is less than or equal to
$\geq$	is greater than or equal to
x^n	nth power of x
$\sqrt{x}$	square root of x
%	percent
$a{:}b$, a/b, $\frac{a}{b}$	ratio of a to b or a divided by b or the fraction $\frac{a}{b}$
$^{\circ}$	degree
(a,b)	ordered pair
$\overrightarrow{AS}$	line AS
$\overline{AS}$	line segment AS
$\overrightarrow{AS}$	ray AS
∟	right angle
$\perp$	is perpendicular to
$\parallel$	is parallel to
$\triangle ABC$	triangle ABC
$\angle ABC$	angle ABC
$\angle B$	angle B

Reference

Latitude and Longitude

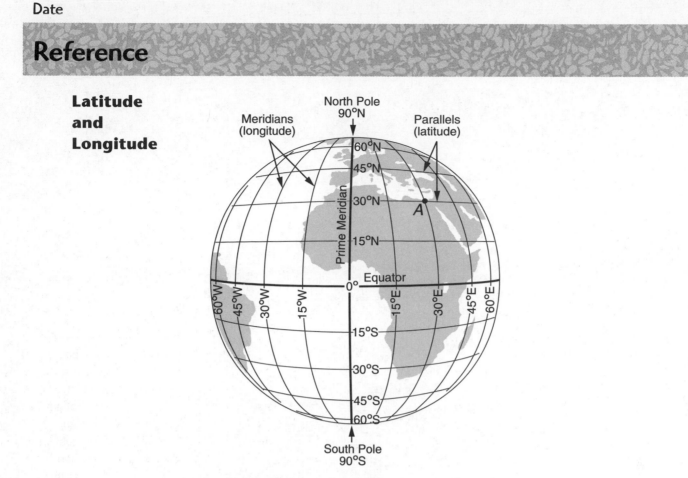

Point *A* is located at 30°N latitude and 30°E longitude.

Fraction-Stick and Decimal Number-Line Chart

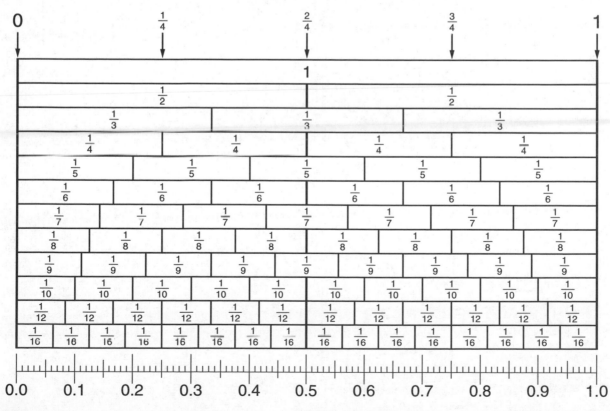

Polygon Capture Property Cards (Front)

There is only one right angle.	There are one or more right angles.	All angles are right angles.	There are no right angles.
There is at least one acute angle.	At least one angle is more than 90°.	All angles are right angles.	There are no right angles.
All opposite sides are parallel.	Only one pair of sides is parallel.	There are no parallel sides.	All sides are the same length.
All opposite sides are parallel.	Some sides have the same length.	All opposite sides have the same length.	**Wild Card:** Pick your own side property.

Angles	Angles	Angles	Angles
Angles	Angles	Angles	Angles
Sides	Sides	Sides	Sides
Sides	Sides	Sides	Sides

Name _____

Date   Time

Slide Rule

Assembly Instructions

1. Cut along the solid lines.

2. Score and fold along the dashed line of the holder so that the number lines are on the outside.

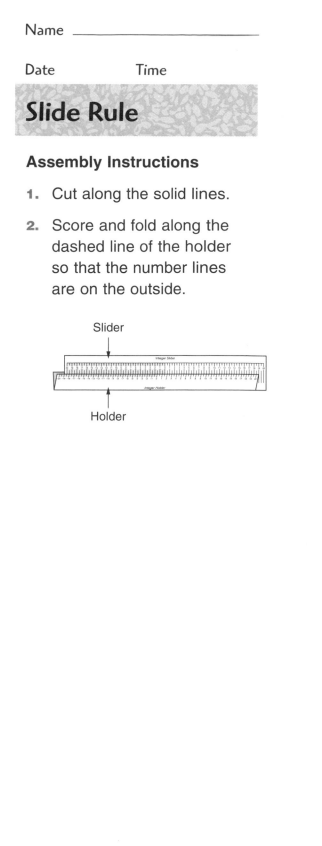

Slider

Holder

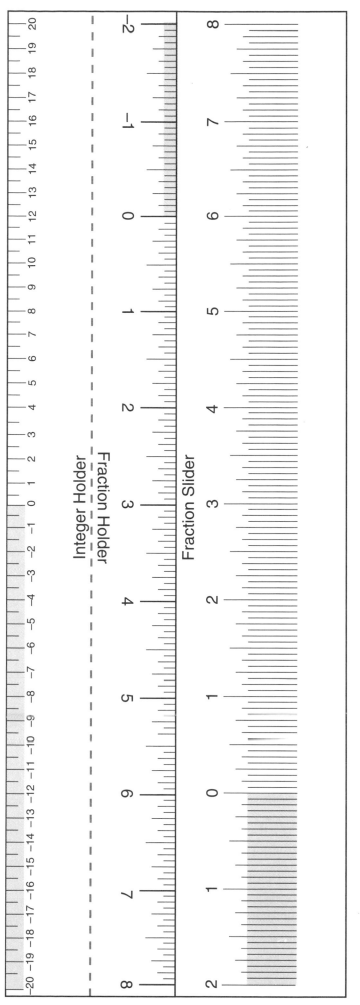

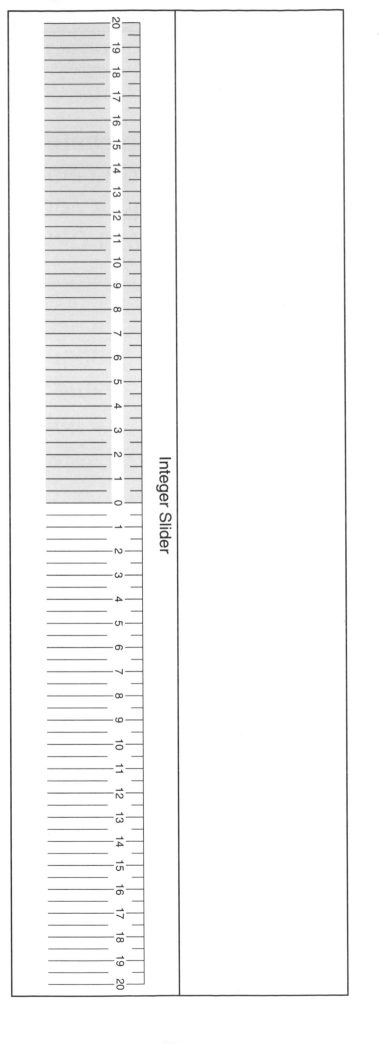

Integer Slider

Rectangular Prism Patterns

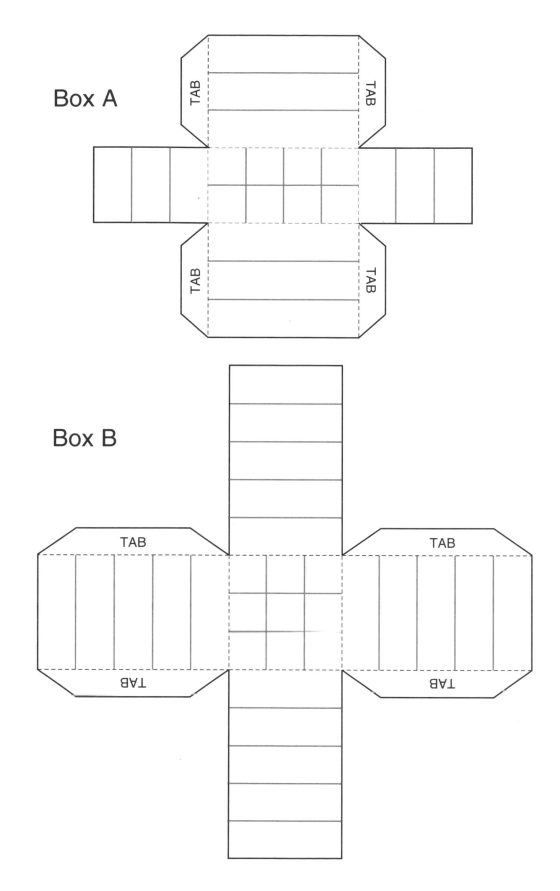

Box A

Box B